W0043614

Akira Miyazaki, Michio Imawari (Eds.)

New Frontiers in Lifestyle-Related Diseases

Akira Miyazaki,
Michio Imawari (Eds.)

New Frontiers in Lifestyle-Related Diseases

 Springer

Akira Miyazaki, M.D., Ph.D.
Professor and Chairman
Department of Biochemistry
Showa University School of Medicine
1-5-8 Hatanodai, Shinagawa-ku, Tokyo 142-8555, Japan

Michio Imawari, M.D., D.M.Sc.
Professor and Chairman
Second Department of Internal Medicine
Showa University School of Medicine
1-5-8 Hatanodai, Shinagawa-ku, Tokyo 142-8555, Japan

Library of Congress Control Number: 2007939280

ISBN   978-4-431-76427-4

This work is subject to copyright. All rights are reserved, whether the whole or part of the material is concerned, specifically the rights of translation, reprinting, reuse of illustrations, recitation, broadcasting, reproduction on microfilms or in other ways, and storage in data banks.
The use of registered names, trademarks, etc. in this publication does not imply, even in the absence of a specific statement, that such names are exempt from the relevant protective laws and regulations and therefore free for general use.
Product liability: The publisher can give no guarantee for information about drug dosage and application thereof contained in this book. In every individual case the respective user must check its accuracy by consulting other pharmaceutical literature.

Springer is a part of Springer Science+Business Media
springer.com
© Springer 2008

Typesetting: Camera-ready by the editors and authors

Printed on acid-free paper

Foreword

It is my great pleasure to publish *New Frontiers in Lifestyle-Related Diseases,* the proceedings of the Showa University International Symposium for Life Sciences, 3rd Annual Meeting, held at Showa University on September 13, 2006. This symposium was supported, in part, by Grants for the Promotion of the Advancement of Education and Research in Graduate Schools and Ordinary Expenses for Private Schools from the Ministry of Education, Culture, Sports, Science, and Technology of Japan. On behalf of Showa University, I would like to express my deepest thanks to all the authors and editors for their great contribution to the publication of this memorable book that accelerates research activity in lifestyle-related diseases.

Akiyoshi Hosoyamada, M.D., Ph.D.
President, Showa University
Tokyo, Japan
September 2007

Preface

The leading cause of death in Western countries and some developing countries is atherosclerotic cardiovascular diseases. Among them, acute myocardial infarction is the most common type of fatal disease, caused by the progression of atherosclerosis characterized by accumulation of cholesterol in vascular walls. Development of atherosclerosis is greatly enhanced by major risk factors for cardiovascular diseases such as obesity, hyperlipidemia, diabetes (hyperglycemia), and hypertension. Among those, obesity frequently initiates a metabolic change that subsequently induces hyperlipidemia, diabetes, hypertension, and eventually atherosclerotic cardiovascular diseases. Because obesity and its related disorders largely depend on lifestyle factors such as high calorie intake and low physical activity, a series of disorders are termed lifestyle-related diseases.

This book includes 14 oral presentations and 5 poster presentations from the Showa University International Symposium, 3rd Annual Meeting, "New Frontiers in Lifestyle-Related Diseases," held on September 13, 2006. The work published here conveys the latest information from basic and clinical research in lifestyle-related diseases. The first part of the book describes regulation of feeding and energy homeostasis. The second part deals with lipid metabolism and atherosclerosis. The third part of the book focuses on novel risk factors for cardiovascular diseases such as sleep apnea syndrome, small dense low-density lipoproteins, and periodontitis.

We wish to express many thanks to all the participants in the symposium and to the authors of this book. We believe this volume provides researchers in the field of lifestyle-related diseases with new and useful information.

Akira Miyazaki, M.D., Ph.D., Professor of Biochemistry
Michio Imawari, M.D., D.M.Sc. Professor of Medicine
Showa University School of Medicine
Tokyo, Japan
September 2007

Contents

Part III. Risk Factors for Cardiovascular Diseases

Part IV. Poster Sessions

Contributors

Akiyama, H., Center for Interdisciplinary Research, Tohoku University, 6-3 Aramaki, Aoba, Sendai, Miyagi 980-8578, Japan

Arai, H., Department of Health Chemistry, Graduate School of Pharmaceutical Sciences, Tokyo University, 7-3-1 Hongo, Bunkyo-ku, Tokyo 113-0033, Japan

Arata, S., Laboratory of DNA Recombination, Showa University, 1-5-8 Hatanodai, Shinagawa-ku, Tokyo 142-8555, Japan

Ban, Y., Third Department of Internal Medicine, Showa University School of Medicine, 1-5-8 Hatanodai, Shinagawa-ku, Tokyo 142-8666, Japan

Chang, C.C.Y., Department of Biochemistry, Dartmouth Medical School, Hanover, NH, 03755 USA

Chang, T.-Y., Department of Biochemistry, Dartmouth Medical School, Hanover, NH, 03755 USA

Civelli, O., Departments of Pharmacology and Developmental and Cell Biology, Schools of Medicine and of Biological Sciences, Med surge II, University California, Irvine, Irvine, CA 92697-4625, USA

Date, Y., Frontier Science Research Center, University of Miyazaki, 5200 Kihara, Kiyotake, Miyazaki 889-1692, Japan

Ge-shi, E., Third Department of Internal Medicine, Showa University School of Medicine, 1-5-8 Hatanodai, Shinagawa-ku, Tokyo 142-8666, Japan

Hirano, T., First Department of Internal Medicine, Showa University School of Medisine, 1-5-8 Hatanodai, Shinagawa-ku, Tokyo 142-8666, Japan

Hongo, S., Department of Biochemistry, Showa University School of Medicine, 1-5-8 Hatanodai, Shinagawa-ku, Tokyo 142-8555, Japan

Itabe, H., Department of Biological Chemistry, School of Pharmaceutical Sciences, Showa University, 1-5-8 Hatanodai, Shinagawa-ku, Tokyo 142-8555, Japan

Kageyama, H., Department of Anatomy I, Showa University School of Medicine, 1-5-8 Hatanodai, Shinagawa-ku, Tokyo 142-8555, Japan

Katagiri, T., Third Department of Internal Medicine, Showa University School of Medicine, 1-5-8 Hatanodai, Shinagawa-ku, Tokyo 142-8666, Japan

Kato, R., Department of Biological Chemistry, School of Pharmaceutical Sciences, Showa University, 1-5-8 Hatanodai, Shinagawa-ku, Tokyo 142-8555, Japan

Kitazato, K., Department of Biological Chemistry, School of Pharmaceutical Sciences, Showa University, 1-5-8 Hatanodai, Shinagawa-ku, Tokyo 142-8555, Japan

Koba, S., Third Department of Internal Medicine, Showa University School of Medicine, 1-5-8 Hatanodai, Shinagawa-ku, Tokyo 142-8666, Japan

Kobayashi, M., Department of Periodontology, Showa University School of Dentistry, 2-1-1 Kitasenzoku, Ohta-ku, Tokyo 145-8515, Japan

Kudo, I., Department of Health Chemistry, School of Pharmaceutical Sciences, Showa University, 1-5-8 Hatanodai, Shinagawa-ku, Tokyo 142-8555, Japan

Kuwata, H., Department of Health Chemistry, School of Pharmaceutical Sciences, Showa University, 1-5-8 Hatanodai, Shinagawa-ku, Tokyo 142-8555, Japan

Masuda, Y., Department of Biological Chemistry, School of Pharmaceutical Sciences, Showa University, 1-5-8 Hatanodai, Shinagawa-ku, Tokyo 142-8555, Japan

Minoguchi, K., First Department of Internal Medicine, Showa University School of Medicine, 1-5-8 Hatanodai, Shinagawa-ku, Tokyo 142-8666, Japan

Minokoshi., Y., National Institutes for Physiological Sciences, 38 Nishigonaka Myodaiji, Okazaki, Aichi 444-8585, Japan

Miyazaki, A., Department of Biochemistry, Showa University School of Medicine, 1-5-8 Hatanodai, Shinagawa-ku, Tokyo 142-8555, Japan

Miyazawa, Y., Department of Periodontology, Showa University School of Dentistry, 2-1-1 Kitasenzoku, Ohta-ku, Tokyo 145-8515, Japan

Mori, C., Department of Biological Chemistry, School of Pharmaceutical Sciences, Showa University, 1-5-8 Hatanodai, Shinagawa-ku, Tokyo 142-8555, Japan

Nagasaki, H., Department of Metabolic Medicine, Nagoya University, School of Medicine, 65 Tsuru-mai, Showa-ku, Nagoya, Aichi 466-8550, Japan

Nakatani, Y., Department of Health Chemistry, School of Pharmaceutical Sciences, Showa University, 1-5-8 Hatanodai, Shinagawa-ku, Tokyo 142-8555, Japan

Nakazato, M., Division of Endocrinology and Metabolism, Department of Internal Medicine, Faculty of Medicine, University of Miyazaki, 5200 Kihara, Kiyotake, Miyazaki 889-1692, Japan

O'Donnell, C. P., Division of Allergy, Pulmonary, and Critical Care Medicine, Department of Medicine, University of Pittsburgh, NW628 MUH 3459 Fifth Ave, Pittsburgh, PA 15213, USA

Obama, T., Department of Biological Chemistry, School of Pharmaceutical Sciences, Showa University, 1-5-8 Hatanodai, Shinagawa-ku, Tokyo 142-8555, Japan

Okamatsu, Y., Department of Periodontology, Showa University School of Dentistry, 2-1-1 Kitasenzoku, Ohta-ku, Tokyo 145-8515, Japan; Dental Clinic, Showa University Hospital, 1-5-8 Hatanodai, Shinagawa-ku, Tokyo 142-8666, Japan

Osaka, T., Division of Endocrinology and Metabolism, Department of Internal Medicine, Faculty of Medicine, University of Miyazaki, 5200 Kihara, Kiyotake, Miyazaki 889-1692, Japan

Saito, Y., Laboratory for Behavioral Neuroscience, Graduate School of Integrated Arts and Sciences Hiroshima University, 1-7-1 Kagamiyama, Higashi-hiroshima, Hiroshima 739-8521, Japan

Sasabe, N., Department of Biological Chemistry, School of Pharmaceutical Sciences, Showa University, 1-5-8 Hatanodai, Shinagawa-ku, Tokyo 142-8555, Japan

Sato, T., Third Department of Internal Medicine, Showa University School of Medicine, 1-5-8 Hatanodai, Shinagawa-ku, Tokyo 142-8666, Japan

Shimazaki, K., Department of Physiological Chemistry, School of Pharmaceutical Sciences, Showa University, 1-5-8 Hatanodai, Shinagawa-ku, Tokyo 142-8555, Japan

Shioda, S., Department of Anatomy I, Showa University School of Medicine, 1-5-8 Hatanodai, Shinagawa-ku, Tokyo 142-8555, Japan

Shoji, M., Third Department of Internal Medicine, Showa University School of Medicine, 1-5-8 Hatanodai, Shinagawa-ku, Tokyo 142-8666, Japan

Suda, R., Department of Periodontology, Showa University School of Dentistry, 2-1-1 Kitasenzoku, Ohta-ku, Tokyo 145-8515, Japan

Suzuki, H., Third Department of Internal Medicine, Showa University School of Medicine, 1-5-8 Hatanodai, Shinagawa-ku, Tokyo 142-8666, Japan

Suzuki, M., Department of Periodontology, Showa University School of Dentistry, 2-1-1 Kitasenzoku, Ohta-ku, Tokyo 145-8515, Japan

Takahashi, K., Department of Biological Chemistry, School of Pharmaceutical Science, Showa University, 1-5-8 Hatanodai, Shinagawa-ku, Tokyo 142-8555, Japan

Takano, T., Department of Molecular Pathology, Faculty of Pharmaceutical Sciences, Teikyo University, 1091-1 Suarashi, Sagamihara, Kanagawa 199-0195, Japan

Takenoya, F., Department of Physical Education, Hoshi University School of Pharmacy and Pharmaceutical Science, 2-4-41 Ebara, Shinagawa-ku, Tokyo 142-8501, Japan

Takiguchi, H., Department of Periodontology, Showa University School of Dentistry, 2-1-1 Kitasenzoku, Ohta-ku, Tokyo 145-8515, Japan

Toshinai, K., Division of Endocrinology and Metabolism, Department of Internal Medicine, Faculty of Medicine, University of Miyazaki, 5200 Kihara, Kiyotake, Miyazaki 889-1692, Japan

Tsunoda, F., Third Department of Internal Medicine, Showa University School of Medicine, 1-5-8 Hatanodai, Shinagawa-ku, Tokyo 142-8666, Japan

Uchida, C., Center for Interdisciplinary Research, Tohoku University, 6-3 Aramaki, Aoba, Sendai, Miyagi 980-8578, Japan

Uchida, T., Center for Interdisciplinary Research, Tohoku University, 6-3 Aramaki, Aoba, Sendai, Miyagi 980-8578, Japan

Ueno, H., Neurology, Respirology, Endocrinology and Metabolism, Miyazaki Medical College, University of Miyazaki, 5200 Kiyotake, Miyazaki 889-1692, Japan

Usui, M., Department of Periodontology, Showa University School of Dentistry, 2-1-1 Kitasenzoku, Ohta-ku, Tokyo 145-8515, Japan

Watanabe, T., Department of Biochemistry, Showa University School of Medicine, 1-5-8 Hatanodai, Shinagawa-ku, Tokyo 142-8555, Japan

Xu, Y., Department of Psychiatry and Behavioral Sciences, Stanford University School of Medicine, 701B Welch Rd Palo Alto, CA 94304-5742, USA

Yamamoto, M., Department of Periodontology, Showa University School of Dentistry, 2-1-1 Kitasenzoku, Ohta-ku, Tokyo 145-8515, Japan

Yokota, Y., Third Department of Internal Medicine, Showa University School of Medicine, 1-5-8 Hatanodai, Shinagawa-ku, Tokyo 142-8666, Japan

Color Plates

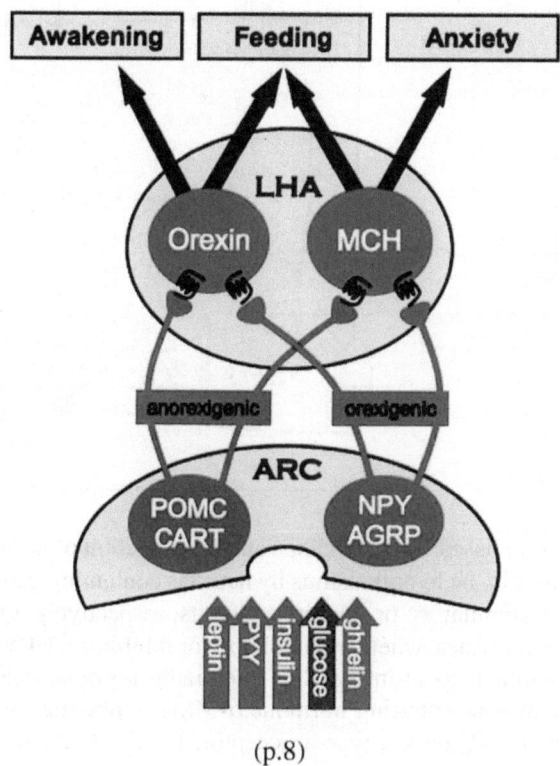

(p.8)

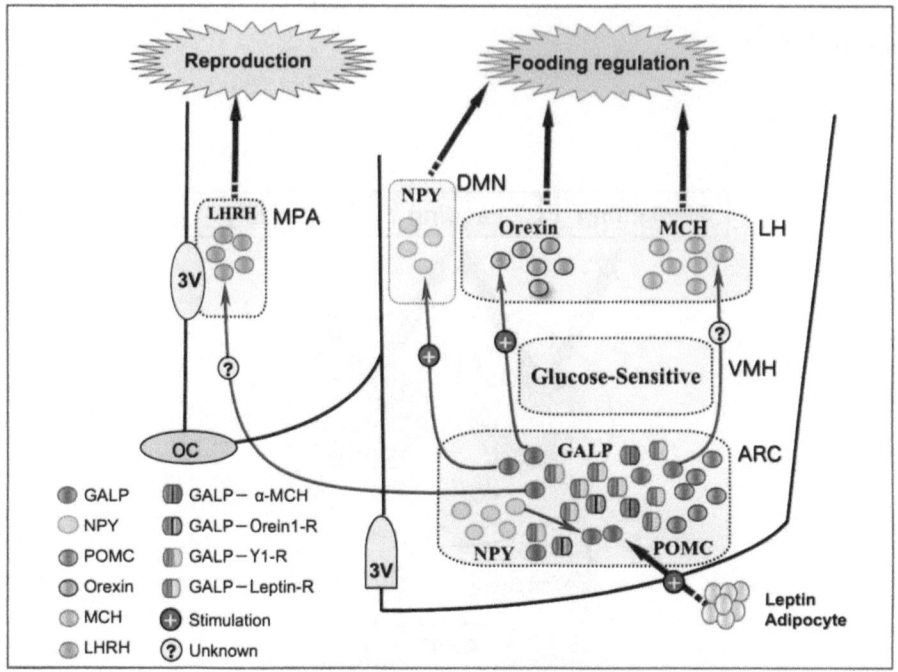

Schematic illustration based on the findings and reproduction of morphological and physiological studies in the hypothalamus by neurons containing peptides. The plus or minus indicates stimulatory or inhibitory effects, respectively. Question marks indicate still unsolved issues, whether stimulatory or inhibitory. NPY; neuropeptide Y, POMC; pro-opiomelanocortin, LHRH; luteinizing hormone-releasing hormonem, MCH: melanin-concentrating hormone,α-MSH; alpha-melanocyte-stimulating hormone, Orexin1-R; orexin type 1 receptor, Leptin-R; leptin receptor, 3 V; third ventricle. (p.35)

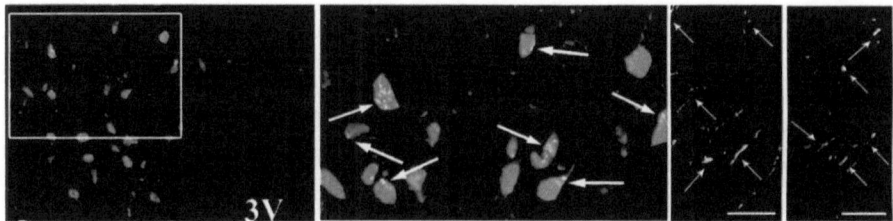

Immunofluorescence photomicrographs of sections with direct dual-labeling combining antisera to GALP and LHRH. A; Immunoreactivity for GALP (red) and LHRH (green) is localized in the MPA respectively. B; The merged images show some GALP-positive terminals in direct contact (arrows) with LHRH neuronal cell bodies and processes. C; Immunoreactivity for GALP (red) and LHRH (green) is localized in the BST. Some GALP-positive terminals in direct contact with LHRH neuronal processes (arrows). Scale bars = 20 μm. 3 V: third ventricle. (p.51)

Cholesterol Trafficking & Esterification

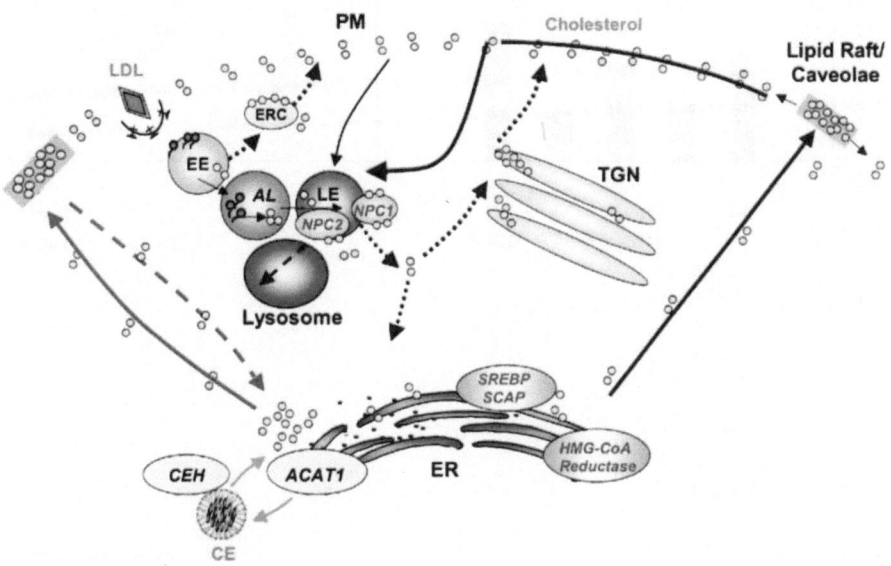

Light yellow circles represent cholesterol molecules. This model is an extension and revision of the earlier model drawn in (Chang et al. 2006). The plasma membranes (PMs) contain the highest concentration of cholesterol. The cholesterol-sensing membrane proteins are located in the ER [HMG-CoA reductase (HMGR), SREBP cleavage–activating protein (SCAP), and acyl-coenzyme A:cholesterol acyltransferase 1 (ACAT1)] or in the late endosomes [Niemann-Pick type C1 (NPC1)]. The translocation of cholesterol between various compartments may involve both vesicular and nonvesicular mechanisms. The dotted lines represent cholesterol trafficking steps that are not well documented. Other abbreviations used: AL, acid lipase; CEH, cholesteryl ester hydrolase; EE, early endosome; ERC, endocytic recycling compartment; LE, late endosome; NPC2, Niemann-Pick type C2; SREBP, sterol-regulatory element–binding protein; TGN, trans-Golgi network. Refer to color plates. (p.65)

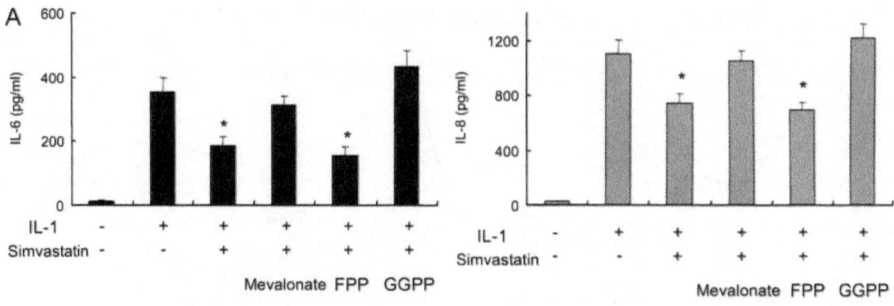

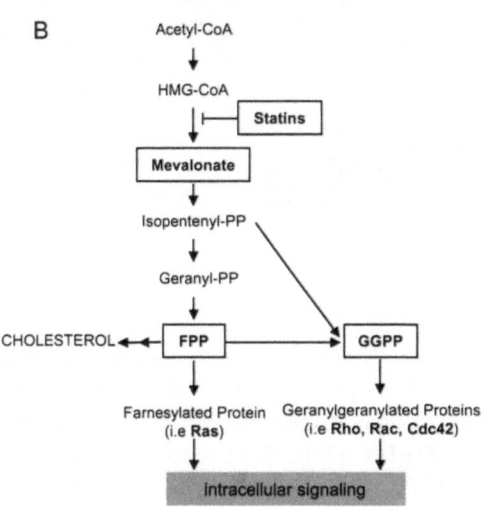

Reversal of inhibitory effect of simvastatin on epthelial cells by co-treatment with downstream metabolites of HMG-CoA reductase.(A) Inhibitory effects of simvastatin (10^{-6} M) on KB cells were reversed by co-treatment with mevalonate (10^{-4} M) or GGPP (5×10^{-6} M), but not with FPP (5×10^{-6} M). Data are expressed as means $\pm$ SD (n = 4). * P < 0.001 vs. control (treated with IL-1) (B)Schematic representation of the mevalonate pathway. Statins block conversion of HMG-CoA to mevalonate. This leads to reduced synthesis of cholesterol and decreased prenylation of proteins such as small GTPases. Isopentenyl-PP, isopentenyl pyrophosphate; Geranyl-PP, geranyl pyrophosphate. (p.129)

Part I
Regulation of Feeding and Energy Homeostasis in the Brain

Orphan Neuropeptides and the Regulation of Food Intake

Hiroshi Nagasaki[1], Yanling Xu[2], Yumiko Saito[3] and Olivier Civelli[4]*

[1]Department of Metabolic Medicine, Nagoya University, School of Medicine, 65 Tsuru-mai, Showa-ku, Nagoya City, Aichi 466-8550, Japan
[2]Department of Psychiatry and Behavioral Sciences, Stanford University School of Medicine, 701B Welch Rd Palo Alto, CA 94304-5742
[3]Laboratory for Behavioral Neuroscience, Graduate School of Integrated Arts and Sciences Hiroshima University, 1-7-1 Kagamiyama, Higashi-hiroshima, Hiroshima 739-8521, Japan
[4]Departments of Pharmacology and Developmental and Cell Biology, Schools of Medicine and of Biological Sciences, Med surge II, University California, Irvine, Irvine, CA 92697-4625

Summary. The completion of the human genome project has led to identification of approximately 800 genes belonging to the G protein-coupled receptor (GPCR) superfamily. Among these are a number of "orphan" GPCRs, receptors with no known endogenous ligands. Although orphan GPCRs are genes without functions, they offer the potential to discover new intercellular interactions and new insights for basic research and ultimately for drug discovery, but their endogenous ligands must be identified for this to happen. The deorphanization of GPCRs has been an ongoing effort since the first GPCRs were cloned by homology approaches. More recently, orphan GPCRs have been used as targets to identify novel endogenous transmitters. This strategy found several neuropeptides. Two are the subjects of this chapter: melanin-concentrating hormone (MCH) and the orexins. Both of these peptides were found to be expressed in the lateral hypothalamic nucleus, one of the brain center regulating food intake. They were studied for their potential role in food intake, and indeed, were both found to exhibit orexigenic activity.

*To whom correspondence should be sent
Supported by grants MH60231, DK070619, DK063001, bio05-10485

4

Keywords. orphan GPCRs, orexins, melanin-concentrating hormone, food intake, obesity

1. Introduction

The completion of the human genome project led to identification of approximately 800 genes belonging to the G protein-coupled receptor (GPCR) superfamily. Among these are a number of "orphan" GPCRs, receptors with no known endogenous ligands. Although orphan GPCRs are genes without functions, they offer the potential to discover new intercellular interactions and new insights for basic research and ultimately for drug discovery, but their endogenous ligands must be identified for this to happen. The deorphanization of GPCRs has been an ongoing effort since the first GPCRs were cloned by homology approaches. More recently, orphan GPCRs have been used as targets to identify novel endogenous transmitters. Since 1995, varieties of bioactive materials including peptides, trace-amine, fatty acids, ADP-libose, cannabinoids have been discovered and these have already greatly enriched our understanding of several physiological responses. Therefore, the deorphanization of GPCR have a great potential to produce seed chemicals on various aspects of pharmacology, including memory, depression, self-administration, endocrine, sleep, energy metabolism, and food intake (Civelli *et al.*, 2001; Douglas *et al.*, 2004).

In this chapter, we discuss the basis of the orphan receptor strategy and its application on the deorphanization of two orphan GPCRs, HFGAN72 and SLC-1, whose biological ligands were found in the lateral hypothalamic nucleus, a brain center in the regulation of feeding.

2. Finding the natural ligands of orphan GPCRs

Most GPCRs have been discovered on the basis of their sequence similarities. Either they were cloned using homology screening approaches or they were discovered by bioinformatic analyses. Noteworthy, the GPCRs discovered in these ways lack natural ligands, they all are "orphan" GPCRs. Finding the endogenous ligands of orphan receptors is a difficult task but one that is primordial if the function of the orphan GPCR is to be studied.

The overall strategy leading to the discovery of the natural ligands of orphan GPCRs followed a unique principle. An orphan GPCR is expressed by transfection into a suitable cell system (mostly mammalian or frog cells) that is then exposed to a variety of naturally occurring molecules that may

Orphan receptor strategy

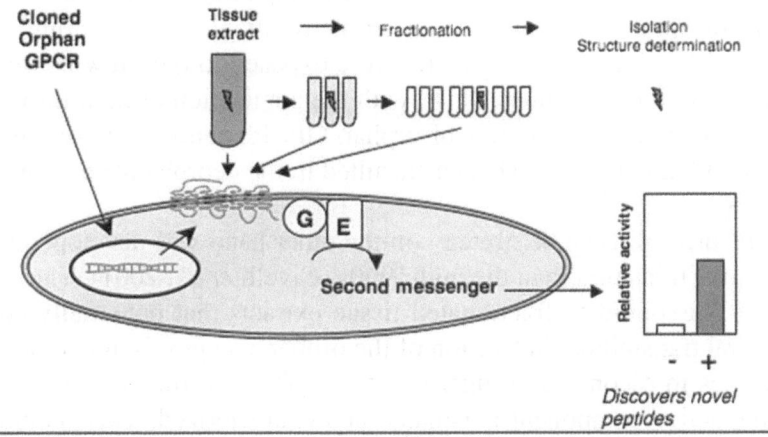

Discovers novel
peptides

Reverse pharmacology

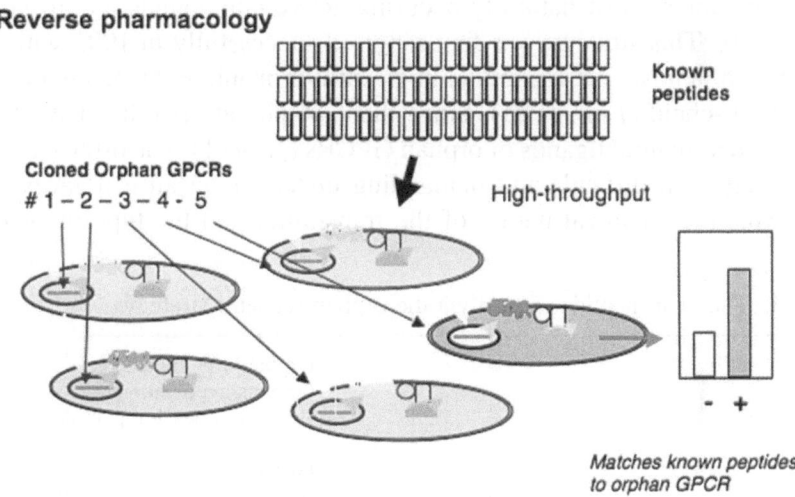

Matches known peptides
to orphan GPCR

Fig. 1.

cause the activation of the GPCR. Orphan GPCR stimulation is monitored through changes in second messenger levels (Figure 1). The presence of the natural ligand is recognized by the change in second messenger levels. Most often, because it is easily applicable to high-throughput screens, measurement of intracellular calcium release has been the parameter of choice in monitoring orphan GPCR reactivity. This approach has been further differentiated into a "reverse pharmacology" in which the orphan GPCR is exposed to known ligands and an "orphan receptor strategy" in which the orphan GPCR is exposed to tissue extracts and is intended to find new transmitters (Civelli *et al.*, 2001).

The reverse pharmacology approach was the first to be applied. It is using this approach that in the late 1980s the dopamine D2 and the serotonin 5HT1A receptors were discovered (Bunzow et al., 1988; Fargin et al., 1988). At that time, orphan GPCRs were exposed to only few neurotransmitters (Civelli et al., 2006). Now, with the application of high-throughput screening techniques, batteries of orphan GPCRs can be exposed to thousands of putative ligands. This has resulted in the deorphanization of dozen of GPCRs.

The orphan receptor strategy on the other hand was developed to discover novel transmitters in the mid-1990s (Civelli et al., 2001). The orphan receptor is exposed to fractionated tissue extracts that potentially contain the natural transmitter. Activation of the orphan receptor is again measured by changes in second messenger responses. Positive fractions are purified until the active component is isolated and characterized. This should lead to the identification of naturally-occuring active compounds, i.e. to novel transmitters. This strategy was first reported successfully in 1995 with the discovery of a novel neuropeptide: nociceptin/orphanin FQ (Meunier et al., 1995; Reinscheid et al., 1995). Since then, 14 peptide families have been discovered as natural ligands of orphan GPCRs (Table 1). The orphan receptor strategy is a difficult and demanding endeavor because it faces two unknowns: the chemical nature of the transmitter and the type of second

Table 1. The neuropeptides found via the orphan receptor strategy

1995	Orphanin FQ/nociceptin[a,b]
1998	Orexins/hypocretins[c,d]
	Prolactin-releasing peptide[e]
	Apelin[f]
1999	Ghrelin[g]
	Melanin-concentrating hormone[h,i,j,k.]
	Urotensin II[l,m,]
2000	Neuromedin U[n,o]
2001	Metastin[p]
2002	Neuropeptide B/Neuropeptide W[q,r,]
	Prokineticin 1–2[s,t]
2003	Relaxin 3[u,v,]
2004	Neuropeptide S[w]
2005	Neuromedin S[x]

[a] (Meunier et al., 1995), [b] (Reinscheid et al., 1995), [c] (de Lecea et al., 1998), [d] (Sakurai et al., 1998b), [e] (Hinuma et al., 1998), [f] (Tatemoto et al., 1998), [g] (Kojima et al., 1999), [h] (Chambers et al., 2000), [i] (Saito et al., 1999), [j] (Shimomura et al., 1999), [k] (Bachner et al., 1999), [l] (Nothacker et al., 1999), [m] (Ames et al., 1999), [n] (Kojima et al., 2000), [o] (Howard et al., 2000), [p] (Ohtaki et al., 2001), [q] (Shimomura et al., 2002), [r] (Tanaka et al., 2003), [s] (LeCouter et al., 2002), [t] (Cheng et al., 2002), [u] (Liu et al., 2003a), [v] (Liu et al., 2003b), [w] (Xu et al., 2004), [x] (Mori et al., 2005)

messenger response that the orphan GPCR will induce. These unknowns, added to the fact that any mammalian cell line contains a battery of endogenous active receptors, make application of the orphan receptor strategy a process requiring careful control and constant adjustment.

3. Two orphan GPCRs associated with the regulation of food intake

Our understanding of the central mechanisms that regulate food intake stems from lesion experiments done in the 1950's (Hetherington, 1940; Anand and Brobeck, 1951). At that time it was shown that destruction of the periventricular nucleus (PVN) initiates feeding, while lesions of lateral hypothalamic area (LH) stops feeding. The PVN became known as a satiety center, and the LH as a hunger center. The studies on the central regulation of feeding did not attract much interest until the 1990s', when the epidemic of the obesity-associated diseases was recognized all over the world. The central regulation of energy homeostasis became a hot research field and a few neuropeptides took the central stage in directing this regulation. Most of these neuronpeptides are ligands of orphan GPCRs. Two of them are discussed in this chapter.

3.1 The orexin system

In 1998, Sakurai et al. deorphanized the orphan HFGAN72 receptor and found that it binds two of closely related peptides, orexin A and B(Sakurai et al., 1998b). They are 33 and 28 amino acids in length respectively, and form a distinctive family from any other known peptides(Sakurai et al., 1998b; Sakurai et al., 1999). They have also been simultaneously discovered as hypothalamic specific peptides and named hypocretins(de Lecea et al., 1998). In mammals, orexins activates two receptor subtypes, OX1R and OX2R(Sakurai et al., 1998b). Orexin neurons are exclusively present in the LH and surrounding regions, and innervates diverse CNS nuclei including the arcuate, locus ceruleus, raphe and tubelomammillary nuclei (Marcus et al., 2001). The distribution of the OXRs corresponds well to the innervation of orexin neurons, although OX1R and OX2R exhibit distinct localizations. OX1R is localized to the hippocampus, amygdala, cerebral cortex, thalamus, hypothalamus, raphe nuclei, locus coeruleus of the midbrain. OX2R is found in the cerebral cortex, hypothalamus, raphe nuclei, laterodorsal tegmental nucleus, pedunculopontine tegmental nucleus. OX2R is dominantly expressed in the central feeding centers such as the arcuate

8

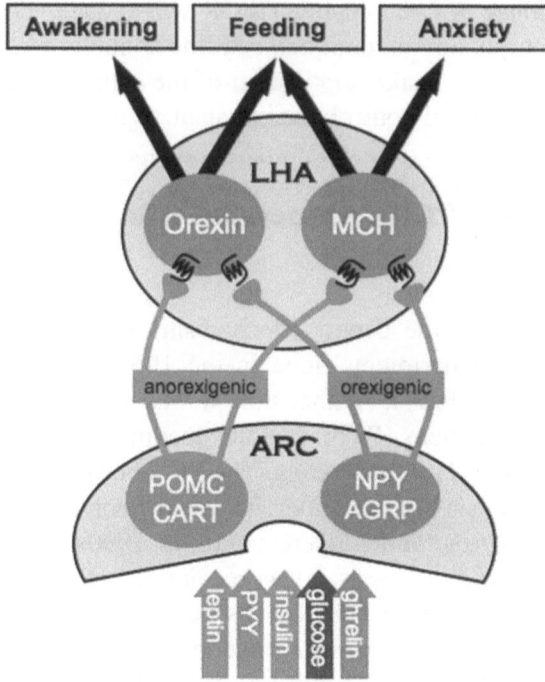

Fig. 2.

nucleus, ventromedial nucleus and lateral hypothalamus. The highest expression of OX1R is found in the locus coeruleus where monoaminergic neurons originate, while the highest expression of OX2R is found in the tuberomammilary nucleus where histaminergic neurons originate. Studies of the innervations of orexin neurons point at the biological roles of orexin system. It regulates feeding through its innervations to hypothalamic neurons including NPY, and also regulates wakefulness through its innervations to the cerebral cortex and limbic regions (Fig 2).

Acute central administration of orexin A stimulates food intake as do injections in the paraventricular hypothalamic nucleus, dorsomedial hypothalamic nucleus, lateral hypothalamus or the perifornical area (Sakurai et al., 1998a; Dube et al., 1999; Sweet et al., 1999). It has been reported that the orexin A effects are circadian specific (Haynes et al., 1999; Thorpe et al., 2003). Central administration of orexin stimulates food intake only when it is administered during the resting (light) phase of rodents (Yamanaka et al., 1999; Thorpe et al., 2003). Similarly, chronic infusion increases daytime food intake, and suppresses it during the active (night) phase, which may be a secondary effect due to the extra food ingested during daytime.

Chronic orexin A administration however does not significantly change food intake over a 24 hour period nor body weight (Haynes *et al.*, 1999; Yamanaka *et al.*, 1999). The effects of orexin B on food intake have been less consistent. Central administration of orexin B has been reported to either have no effect (Haynes *et al.*, 1999) or to stimulate food intake (Sakurai et al, 1998). Site specific microinjections of orexin B in discrete areas of the hypothalamus have failed to demonstrate significant effects on food intake (Dube *et al.*, 1999).

Consistent with the hyperphagic effects of orexin A, prepro-orexin knockout mice exhibit hypophagia while maintaining normal body weight (Willie *et al.*, 2001). Similarly, the selective deletion of orexin neurons using ataxin-3 genetic constructs induces hypophagia in mice which also develop late-onset obesity in contrast to the prepro-orexin knockouts (Hara *et al.*, 2001). This phenotype occurs in spite of hypophagia and may be due to a lower level of energy expenditure that could override the hypophagia and lead to an increase in body weight. This is in agreement with other reports that central administration of orexin A increases oxygen consumption and body temperature, important indexes of energy expenditure (Wang *et al.*, 2001, 2003).

Normal body weight after chronic administration of orexin and late-onset of obesity in mice with ablated orexin neurons indicate that the orexin system may not be the cause of obesity. Studies have reported that prepro-orexin mRNA expression is downregulated in two different obesity animal models with dysfunctional leptin systems (ob/ob and db/db mice) (Yamamoto *et al.*, 1999), as well as in obese Zucker rats (Cai *et al.*, 1999; Beck *et al.*, 2001). Furthermore, fasting increased prepro-orexin mRNA expression (Sakurai *et al.*, 1998a). This suggests that the orexin system may be a feedback regulatory system turned on in response to obesity.

Among the two orexin receptors, the orexin-1 receptor is clearly involved in the regulation of food intake and energy expenditure. The selective orexin-1 receptor antagonist, SB-334867, reduces food intake when given at the start of the dark phase and decreases stimulated feeding induced by overnight fasting in rats. It also blocks the orexin-induced hyperphagia (Haynes *et al.*, 2000). In ob/ob mice, SB-334867 administration leads to the reduction of weight gain with decreased total fat mass gain but no changes in fat free mass. The reduced weight gain is attributed to cumulative food intake and possible stimulation of thermogenesis as indicated by decreased brown adipose tissue (Haynes *et al.*, 2002). The lack of a selective antagonist to the orexin-2 receptor has impeded the study of the specific involvement of the orexin-2 receptor in feeding.

In addition to its modulatory role on food intake, the orexin system is a major regulator of sleep and arousal (Sutcliffe and de Lecea, 2002). Central administration of orexin A increases locomotor activity and promotes wakefulness (Hagan *et al.*, 1999; Bourgin *et al.*, 2000). Orexin knockout and orexin/ataxin-3 transgenic mice are narcoleptic (Chemelli *et al.*, 1999; Hara *et al.*, 2001). Moreover, in humans, significant decreases in the number of orexin containing neurons have been found in postmortem brains of narcoleptic individuals (Thannickal *et al.*, 2000). Among the two receptors, the orexin-2 receptor is the predominant one that is involved in mediating the arousal effects of orexin. Narcolepsy results from mutations that abrogate the activity of the orexin-2 receptor in a canine model (Lin *et al.*, 1999). Orexin-2 receptor knockout mice also are narcoleptic. On the other hand, the orexin-1 receptor knockout mice exhibit only mild fragmentation of the sleep-wake cycle but no obvious narcolepsy (Willie *et al.*, 2003). Together, these data demonstrate the importance of the orexin system in the pathophysiology of narcolepsy and the relative preference of the orexin-2 receptor in mediating the arousal effects of orexin when compared to the orexin-1 receptor.

Consequently, the important issue in evaluating the effects of the orexin system on food intake and energy expenditure is whether the stimulating effects of orexin could be due to increased activity or wakefulness. Several data have suggested that the effects of the orexin system on energy balance is not simply the secondary effects on arousal induced by orexin. In anesthetized rats, orexin A still had the ability to increase oxygen consumption, skin and body temperature, which are signs of increased energy expenditure, in spite of the anesthetized condition (Wang *et al.*, 2001). Moreover, orexin-1 receptor antagonists reduce food intake and decrease body weight while the orexin-1 receptor knockout mice induce only a slight fragmentation of the sleep-wake cycle without obvious behavioral abnormality. Interestingly, recent evidence demonstrates that the orexin system plays a role in linking the regulation of energy balance with the regulation of arousal. The orexin system has been shown to modulate food seeking behavior in response to energy balance. Signals involved in metabolic and energy balance such as glucose, leptin and ghrelin, regulate the activity of orexin neurons and expression levels of orexin. Orexin neurons are able to induce or suppress food seeking behavior by integrating energy-related signals coming from peripheral or hypothalamic pathways. It has been shown that mice that lack orexin expressing neurons (orexin/ataxin-3 transgenic mice) fail to increase their wakefulness state and their motor activity in response to fasting (Yamanaka *et al.*, 2003). Therefore the orexin system appears to link adaptive food seeking behavior in response to energy homeostasis to the regulation of arousal (Sakurai, 2003).

3.2 The MCH system

The SLC-1 orphan GPCR was originally discovered as an expressed sequence tag exhibiting about 40% homology in its hydrophobic domains to the five somatostatin receptors(Kolakowski *et al.*, 1996). A rat ortholog that shares 91% overall sequence identity to the human SLC-1 receptor but is 49 amino acids shorter in its N-terminal segment (Lakaye *et al.*, 1998) was later identified. The existence of a shorter human form was also reported. In 1999, five independent groups, including our own, reported the identity of the natural ligand of SLC-1 by using the two different strategies described for the deorphanization of orphan GPCRs(Bachner *et al.*, 1999; Lembo *et al.*, 1999; Saito *et al.*, 1999; Shimomura *et al.*, 1999; Chambers *et al.*, 2000). First, three groups described the successful application of the orphan receptor strategy. They used brain extracts as starting materials and although they reached the same result, they monitored SLC-1 activity via three different second messenger responses (calcium influx with chimeric Gαprotein, cAMP inhibition assay, and G protein-gated potassium channels (Bachner *et al.*, 1999; Saito *et al.*, 1999; Shimomura *et al.*, 1999). Two other groups used the reverse pharmacology approach and screened large libraries of synthetic molecules as potential activators of SLC-1 (Saito *et al.*, 1999; Chambers *et al.*, 2000) and monitored SLC-1 reactivity by calcium influx measurements. Each group reached the same conclusion: SLC-1 binds the known peptide MCH. MCH is a 19-amino acid cyclic peptide that was originally discovered in teleost fish, where it acts as a skin-color-regulating hormone(Kawauchi *et al.*, 1983). MCH in mammals has been implicated in a variety of physiological functions. Recently, however, its central role in the control of feeding has been the center of attention(Pissios and Maratos-Flier, 2003) (Fig 2).

MCH injections into the lateral ventricles of rats led to increased food intake(Qu *et al.*, 1996; Rossi *et al.*, 1997). MCH knockout mice are hypophagic and exhibit increased metabolic rates, resulting in decreased body weight and body fat content (Shimada *et al.*, 1998). Conversely, transgenic mice overexpressing MCH show a tendency to gain more weight than their wild-type counterparts, displaying high levels of leptin, glucose, and insulin(Ludwig *et al.*, 2001). Another finding is that the MCH system is regulated by energy homeostasis and interacts with other energy metabolic modulators such as leptin. Prepro-MCH mRNA is upregulated both in fasted animals and in leptin-deficient *ob/ob* mice. The MCH system also plays a key role as a downstream mediator of leptin action and is required for the obesity of leptin deficiency. Crossing MCH knockout mice with mice with obesity due to genetic leptin deficiency (*ob/ob*) attenuates phenotypic manifestations of leptin deficiency(Segal-Lieberman *et al.*, 2003). The marked reduction in weight in these double null mice is secondary to decreased total fat body fat. The mice display significantly increased locomotor activity and thermoregulation as compared to *ob/ob* mice but are surprisingly more hyperphagic than *ob/ob* mice. This indicates that the weight loss induced by the absence of MCH in *ob/ob* mice is resulted in increased energy expenditures. Thus, the importance of the MCH system as a potential candidate in obesity treatment has been well documented. However, the system could not be targeted in drug discovery due to the lack of adequate receptor binding assays. The deorphanization of SLC-1 changed this; the MCH system could be studied from both the ligand and receptor standpoints.

The exogenously expressed MCH-1 receptor (MCH-1R) couples with various second messenger systems, including elevation of intracellular Ca^{2+} levels, inhibition of forskolin-stimulated cyclic AMP production, and activation of the mitogen-activated protein kinase cascade(Lembo *et al.*, 1999; Saito *et al.*, 1999; Chambers *et al.*, 2000; Hawes *et al.*, 2000). Whereas it was known that MCH is prominently expressed only in two brain areas known to be involved in feeding behavior—the perikarya of the lateral hypothalamus and the zona incerta(Bittencourt *et al.*, 1992)—the MCH-1 receptor is widely distributed in the central nervous system and is found in particular in regions involved in rewarding behavior, feeding behavior, and metabolic regulation, for example, the arcuate nucleus, ventromedial hypothalamic nucleus, and nucleus accumbens in rats(Lembo *et al.*, 1999; Saito *et al.*, 2001).

Although orexin neurons innervate quite the same regions as the MCH neurons, they are distinct from each others. Indeed it has been shown that MCH receptors cover practically the sum of the regions covered by OX1R and OX2R. On the other hand, orexin and MCH neurons receive input of various neuronal and humoral inputs including leptin, NPY, melanocortin, ghrelin, and glucose. These two systems are therefore viewed as the recipients of the cascade pathway that direct the central regulation of food intake and that originates in the arcuate nucleus with the POMC- and NPY-releasing neurons.

4. Conclusions

The orphan receptor strategy has been successful over the past ten years in identifying a dozen novel neuropeptide families. Interestingly, while these novel neuropeptides have been shown to modulate distinct physiological responses, several have been demonstrated to be important regulators of food intake and energy balance. Why is food intake such a preferential target for newly described peptides over other physiological responses? One reason may be that these peptides have been discovered at the time when our understanding of food intake took a new turn with the discovery of leptin and thus testing of novel natural entities for their effects on food intake was common. This would mean that these novel neuropeptides may have several other physiological effects, some that may even be more pronounced than their effects on obesity, as indeed already demonstrated by the effect of orexins on wakefulness. But it remains true that these peptides have greatly enhanced our understanding of the pathophysiology of obesity. Moreover, being ligands of GPCRs they represent promising targets for developing drugs to treat obesity, a disorder that affects the lives of millions of people.

References

Ames, R.S., Sarau, H.M., Chambers, J.K., Willette, R.N., Aiyar, N.V., Romanic, A.M., Louden, C.S., Foley, J.J., Sauermelch, C.F., Coatney, R.W., Ao, Z., Disa, J., Holmes, S.D., Stadel, J.M., Martin, J.D., Liu, W.S., Glover, G.I., Wilson, S., McNulty, D.E., Ellis, C.E., Elshourbagy, N.A., Shabon, U., Trill, J.J., Hay, D.W., Ohlstein, E.H., Bergsma, D.J., and Douglas, S.A. (1999). Human urotensin-II is a potent vasoconstrictor and agonist for the orphan receptor GPR14. Nature *401*, 282–286.

Anand, B.K., and Brobeck, J.R. (1951). Localization of a "feeding center" in the hypothalamus of the rat. Proc Soc Exp Biol Med *77*, 323–324.

Bachner, D., Kreienkamp, H., Weise, C., Buck, F., and Richter, D. (1999). Identification of melanin concentrating hormone (MCH) as the natural ligand for the orphan somatostatin-like receptor 1 (SLC-1). FEBS Lett *457*, 522–524.

Beck, B., Richy, S., Dimitrov, T., and Stricker-Krongrad, A. (2001). Opposite regulation of hypothalamic orexin and neuropeptide Y receptors and peptide expressions in obese Zucker rats. Biochem Biophys Res Commun *286*, 518–523.

Bittencourt, J.C., Presse, F., Arias, C., Peto, C., Vaughan, J., Nahon, J.L., Vale, W., and Sawchenko, P.E. (1992). The melanin-concentrating hormone system of the rat brain: an immuno- and hybridization histochemical characterization. J Comp Neurol *319*, 218–245.

Bourgin, P., Huitron-Resendiz, S., Spier, A.D., Fabre, V., Morte, B., Criado, J.R., Sutcliffe, J.G., Henriksen, S.J., and de Lecea, L. (2000). Hypocretin-1 modulates rapid eye movement sleep through activation of locus coeruleus neurons. J Neurosci *20*, 7760–7765.

Bunzow, J.R., Van Tol, H.H., Grandy, D.K., Albert, P., Salon, J., Christie, M., Machida, C.A., Neve, K.A., and Civelli, O. (1988). Cloning and expression of a rat D2 dopamine receptor cDNA. Nature *336*, 783–787.

Cai, X.J., Widdowson, P.S., Harrold, J., Wilson, S., Buckingham, R.E., Arch, J.R., Tadayyon, M., Clapham, J.C., Wilding, J., and Williams, G. (1999). Hypotha-

14

lamic orexin expression: modulation by blood glucose and feeding. Diabetes *48*, 2132–2137.

Chemelli, R.M., Willie, J.T., Sinton, C.M., Elmquist, J.K., Scammell, T., Lee, C., Richardson, J.A., Williams, S.C., Xiong, Y., Kisanuki, Y., Fitch, T.E., Naka-zato, M., Hammer, R.E., Saper, C.B., and Yanagisawa, M. (1999). Narcolepsy in orexin knockout mice: molecular genetics of sleep regulation. Cell *98*, 437–451.

Cheng, M.Y., Bullock, C.M., Li, C., Lee, A.G., Bermak, J.C., Belluzzi, J., Weaver, D.R., Leslie, F.M., and Zhou, Q.Y. (2002). Prokineticin 2 transmits the behavioural circadian rhythm of the suprachiasmatic nucleus. Nature *417*, 405–410.

Civelli, O., Nothacker, H.P., Saito, Y., Wang, Z., Lin, S.H., and Reinscheid, R.K. (2001). Novel neurotransmitters as natural ligands of orphan G-protein-coupled receptors. Trends Neurosci *24*, 230–237.

Civelli, O., Saito, Y., Wang, Z., Nothacker, HP. and Reinscheid, R. (2006). Orphan GPCRs and their ligands. Pharmacology & Therapeutics *110*, 525–532.

de Lecea, L., Kilduff, T.S., Peyron, C., Gao, X., Foye, P.E., Danielson, P.E., Fuku-hara, C., Battenberg, E.L., Gautvik, V.T., Bartlett, F.S., 2nd, Frankel, W.N., van den Pol, A.N., Bloom, F.E., Gautvik, K.M., and Sutcliffe, J.G. (1998). The hypocretins: hypothalamus-specific peptides with neuroexcitatory activity. Proc Natl Acad Sci U S A *95*, 322–327.

Douglas, S.A., Ohlstein, E.H., and Johns, D.G. (2004). Techniques: Cardiovascular pharmacology and drug discovery in the 21st century. Trends Pharmacol Sci *25*, 225–233.

Dube, M.G., Kalra, S.P., and Kalra, P.S. (1999). Food intake elicited by central administration of orexins/hypocretins: identification of hypothalamic sites of action. Brain Res *842*, 473–477.

Fargin, A., Raymond, J.R., Lohse, M.J., Kobilka, B.K., Caron, M.G., and Lefkow-itz, R.J. (1988). The genomic clone G-21 which resembles a beta-adrenergic receptor sequence encodes the 5-HT1A receptor. Nature *335*, 358–360.

Hagan, J.J., Leslie, R.A., Patel, S., Evans, M.L., Wattam, T.A., Holmes, S., Benham, C.D., Taylor, S.G., Routledge, C., Hemmati, P., Munton, R.P., Ashmeade, T.E., Shah, A.S., Hatcher, J.P., Hatcher, P.D., Jones, D.N., Smith, M.I., Piper, D.C., Hunter, A.J., Porter, R.A., and Upton, N. (1999). Orexin A activates locus coeruleus cell firing and increases arousal in the rat. Proc Natl Acad Sci U S A *96*, 10911–10916.

Hara, J., Beuckmann, C.T., Nambu, T., Willie, J.T., Chemelli, R.M., Sinton, C.M., Sugiyama, F., Yagami, K., Goto, K., Yanagisawa, M., and Sakurai, T. (2001). Genetic ablation of orexin neurons in mice results in narcolepsy, hypophagia, and obesity. Neuron *30*, 345–354.

Hawes, B.E., Kil, E., Green, B., O'Neill, K., Fried, S., and Graziano, M.P. (2000). The melanin-concentrating hormone receptor couples to multiple G proteins to activate diverse intracellular signaling pathways. Endocrinology *141*, 4524–4532.

Haynes, A.C., Chapman, H., Taylor, C., Moore, G.B., Cawthorne, M.A., Tadayyon, M., Clapham, J.C., and Arch, J.R. (2002). Anorectic, thermogenic and antiobesity activity of a selective orexin-1 receptor antagonist in ob/ob mice. Regul Pept *104*, 153–159.

Haynes, A.C., Jackson, B., Chapman, H., Tadayyon, M., Johns, A., Porter, R.A., and Arch, J.R. (2000). A selective orexin-1 receptor antagonist reduces food consumption in male and female rats. Regul Pept *96*, 45–51.

Haynes, A.C., Jackson, B., Overend, P., Buckingham, R.E., Wilson, S., Tadayyon, M., and Arch, J.R. (1999). Effects of single and chronic intracerebroventricular administration of the orexins on feeding in the rat. Peptides *20*, 1099–1105.

Hetherington, A.W., Ranson S.W. (1940). Hypothalamic lesions and adiposity in the rat. Anat. Rec. *78*, 149–172.

Hinuma, S., Habata, Y., Fujii, R., Kawamata, Y., Hosoya, M., Fukusumi, S., Kitada, C., Masuo, Y., Asano, T., Matsumoto, H., Sekiguchi, M., Kurokawa, T., Nishimura, O., Onda, H., and Fujino, M. (1998). A prolactin-releasing peptide in the brain. Nature *393*, 272–276.

Howard, A.D., Wang, R., Pong, S.S., Mellin, T.N., Strack, A., Guan, X.M., Zeng, Z., Williams, D.L., Jr., Feighner, S.D., Nunes, C.N., Murphy, B., Stair, J.N., Yu, H., Jiang, Q., Clements, M.K., Tan, C.P., McKee, K.K., Hreniuk, D.L., McDonald, T.P., Lynch, K.R., Evans, J.F., Austin, C.P., Caskey, C.T., Van der Ploeg, L.H., and Liu, Q. (2000). Identification of receptors for neuromedin U and its role in feeding. Nature *406*, 70–74.

Kawauchi, H., Kawazoe, I., Tsubokawa, M., Kishida, M., and Baker, B.I. (1983). Characterization of melanin-concentrating hormone in chum salmon pituitaries. Nature *305*, 321–323.

Kojima, M., Haruno, R., Nakazato, M., Date, Y., Murakami, N., Hanada, R., Matsuo, H., and Kangawa, K. (2000). Purification and identification of neuromedin U as an endogenous ligand for an orphan receptor GPR66 (FM3). Biochem Biophys Res Commun *276*, 435–438.

Kojima, M., Hosoda, H., Date, Y., Nakazato, M., Matsuo, H., and Kangawa, K. (1999). Ghrelin is a growth-hormone-releasing acylated peptide from stomach. Nature *402*, 656–660.

Kolakowski, L.F., Jr., Jung, B.P., Nguyen, T., Johnson, M.P., Lynch, K.R., Cheng, R., Heng, H.H., George, S.R., and O'Dowd, B.F. (1996). Characterization of a human gene related to genes encoding somatostatin receptors. FEBS Lett *398*, 253–258.

Lakaye, B., Minet, A., Zorzi, W., and Grisar, T. (1998). Cloning of the rat brain cDNA encoding for the SLC-1 G protein-coupled receptor reveals the presence of an intron in the gene. Biochim Biophys Acta *1401*, 216–220.

LeCouter, J., Lin, R., and Ferrara, N. (2002). Endocrine gland-derived VEGF and the emerging hypothesis of organ-specific regulation of angiogenesis. Nat Med *8*, 913–917.

Lembo, P.M., Grazzini, E., Cao, J., Hubatsch, D.A., Pelletier, M., Hoffert, C., St-Onge, S., Pou, C., Labrecque, J., Groblewski, T., O'Donnell, D., Payza, K., Ahmad, S., and Walker, P. (1999). The receptor for the orexigenic peptide melanin-concentrating hormone is a G-protein-coupled receptor. Nat Cell Biol *1*, 267–271.

Libert, F., Parmentier, M., Lefort, A., Dinsart, C., Van Sande, J., Maenhaut, C., Simons, M.J., Dumont, J.E., and Vassart, G. (1989). Selective amplification and cloning of four new members of the G protein-coupled receptor family. Science *244*, 569–572.

Lin, L., Faraco, J., Li, R., Kadotani, H., Rogers, W., Lin, X., Qiu, X., de Jong, P.J., Nishino, S., and Mignot, E. (1999). The sleep disorder canine narcolepsy is caused by a mutation in the hypocretin (orexin) receptor 2 gene. Cell *98*, 365–376.

Liu, C., Chen, J., Sutton, S., Roland, B., Kuei, C., Farmer, N., Sillard, R., and Lovenberg, T.W. (2003a). Identification of relaxin-3/INSL7 as a ligand for GPCR142. J Biol Chem *278*, 50765–50770.

Liu, C., Eriste, E., Sutton, S., Chen, J., Roland, B., Kuei, C., Farmer, N., Jornvall, H., Sillard, R., and Lovenberg, T.W. (2003b). Identification of relaxin-3/INSL7 as an endogenous ligand for the orphan G-protein-coupled receptor GPCR135. J Biol Chem *278*, 50754–50764.

Ludwig, D.S., Tritos, N.A., Mastaitis, J.W., Kulkarni, R., Kokkotou, E., Elmquist, J., Lowell, B., Flier, J.S., and Maratos-Flier, E. (2001). Melanin-concentrating hormone overexpression in transgenic mice leads to obesity and insulin resistance. J Clin Invest *107*, 379–386.

Marcus, J.N., Aschkenasi, C.J., Lee, C.E., Chemelli, R.M., Saper, C.B., Yanagi-sawa, M., and Elmquist, J.K. (2001). Differential expression of orexin receptors 1 and 2 in the rat brain. J Comp Neurol *435*, 6–25.

Meunier, J.C., Mollereau, C., Toll, L., Suaudeau, C., Moisand, C., Alvinerie, P., Butour, J.L., Guillemot, J.C., Ferrara, P., Monsarrat, B., and et al. (1995). Isolation and structure of the endogenous agonist of opioid receptor-like ORL1 receptor. Nature *377*, 532–535.

Mori, K., Miyazato, M., Ida, T., Murakami, N., Serino, R., Ueta, Y., Kojima, M., and Kangawa, K. (2005). Identification of neuromedin S and its possible role in the mammalian circadian oscillator system. Embo J *24*, 325–335.

Nothacker, H.P., Wang, Z., McNeill, A.M., Saito, Y., Merten, S., O'Dowd, B., Duckles, S.P., and Civelli, O. (1999). Identification of the natural ligand of an orphan G-protein-coupled receptor involved in the regulation of vasoconstriction. Nat Cell Biol *1*, 383–385.

Ohtaki, T., Shintani, Y., Honda, S., Matsumoto, H., Hori, A., Kanehashi, K., Terao, Y., Kumano, S., Takatsu, Y., Masuda, Y., Ishibashi, Y., Watanabe, T., Asada, M., Yamada, T., Suenaga, M., Kitada, C., Usuki, S., Kurokawa, T., Onda, H., Nishimura, O., and Fujino, M. (2001). Metastasis suppressor gene KiSS-1 encodes peptide ligand of a G-protein-coupled receptor. Nature *411*, 613–617.

Pissios, P., and Maratos-Flier, E. (2003). Melanin-concentrating hormone: from fish skin to skinny mammals. Trends Endocrinol Metab *14*, 243–248.

Qu, D., Ludwig, D.S., Gammeltoft, S., Piper, M., Pelleymounter, M.A., Cullen, M.J., Mathes, W.F., Przypek, R., Kanarek, R., and Maratos-Flier, E. (1996). A role for melanin-concentrating hormone in the central regulation of feeding behaviour. Nature *380*, 243–247.

Reinscheid, R.K., Nothacker, H.P., Bourson, A., Ardati, A., Henningsen, R.A., Bunzow, J.R., Grandy, D.K., Langen, H., Monsma, F.J., Jr., and Civelli, O.

(1995). Orphanin FQ: a neuropeptide that activates an opioidlike G protein-coupled receptor. Science *270*, 792–794.

Rossi, M., Choi, S.J., O'Shea, D., Miyoshi, T., Ghatei, M.A., and Bloom, S.R. (1997). Melanin-concentrating hormone acutely stimulates feeding, but chronic administration has no effect on body weight. Endocrinology *138*, 351–355.

Saito, Y., Nothacker, H.P., Wang, Z., Lin, S.H., Leslie, F., and Civelli, O. (1999). Molecular characterization of the melanin-concentrating-hormone receptor. Nature *400*, 265–269.

Saito, Y., Wang, Z., Hagino-Yamagishi, K., Civelli, O., Kawashima, S., and Maruyama, K. (2001). Endogenous melanin-concentrating hormone receptor SLC-1 in human melanoma SK-MEL-37 cells. Biochem Biophys Res Commun *289*, 44–50.

Sakurai, T. (2003). Orexin: a link between energy homeostasis and adaptive behaviour. Curr Opin Clin Nutr Metab Care *6*, 353–360.

Sakurai, T., Amemiya, A., Ishii, M., Matsuzaki, I., Chemelli, R.M., Tanaka, H., Williams, S.C., Richardson, J.A., Kozlowski, G.P., Wilson, S., Arch, J.R., Buckingham, R.E., Haynes, A.C., Carr, S.A., Annan, R.S., McNulty, D.E., Liu, W.S., Terrett, J.A., Elshourbagy, N.A., Bergsma, D.J., and Yanagisawa, M. (1998a). Orexins and orexin receptors: a family of hypothalamic neuropeptides and G protein-coupled receptors that regulate feeding behavior. Cell *92*, 573–585.

Sakurai, T., Amemiya, A., Ishii, M., Matsuzaki, I., Chemelli, R.M., Tanaka, H., Williams, S.C., Richarson, J.A., Kozlowski, G.P., Wilson, S., Arch, J.R., Buckingham, R.E., Haynes, A.C., Carr, S.A., Annan, R.S., McNulty, D.E., Liu, W.S., Terrett, J.A., Elshourbagy, N.A., Bergsma, D.J., and Yanagisawa, M. (1998b). Orexins and orexin receptors: a family of hypothalamic neuropeptides and G protein-coupled receptors that regulate feeding behavior. Cell *92*, 1 page following 696.

Sakurai, T., Moriguchi, T., Furuya, K., Kajiwara, N., Nakamura, T., Yanagisawa, M., and Goto, K. (1999). Structure and function of human prepro-orexin gene. J Biol Chem *274*, 17771–17776.

Segal-Lieberman, G., Bradley, R.L., Kokkotou, E., Carlson, M., Trombly, D.J., Wang, X., Bates, S., Myers, M.G., Jr., Flier, J.S., and Maratos-Flier, E. (2003). Melanin-concentrating hormone is a critical mediator of the leptin-deficient phenotype. Proc Natl Acad Sci U S A *100*, 10085–10090.

Shimada, M., Tritos, N.A., Lowell, B.B., Flier, J.S., and Maratos-Flier, E. (1998). Mice lacking melanin-concentrating hormone are hypophagic and lean. Nature *396*, 670–674.

Shimomura, Y., Harada, M., Goto, M., Sugo, T., Matsumoto, Y., Abe, M., Watanabe, T., Asami, T., Kitada, C., Mori, M., Onda, H., and Fujino, M. (2002). Identification of neuropeptide W as the endogenous ligand for orphan G-protein-coupled receptors GPR7 and GPR8. J Biol Chem *277*, 35826–35832.

Shimomura, Y., Mori, M., Sugo, T., Ishibashi, Y., Abe, M., Kurokawa, T., Onda, H., Nishimura, O., Sumino, Y., and Fujino, M. (1999). Isolation and identification of melanin-concentrating hormone as the endogenous ligand of the SLC-1 receptor. Biochem Biophys Res Commun *261*, 622–626.

Sutcliffe, J.G., and de Lecea, L. (2002). The hypocretins: setting the arousal threshold. Nat Rev Neurosci 3, 339–349.

Sweet, D.C., Levine, A.S., Billington, C.J., and Kotz, C.M. (1999). Feeding response to central orexins. Brain Res 821, 535–538.

Tanaka, H., Yoshida, T., Miyamoto, N., Motoike, T., Kurosu, H., Shibata, K., Yamanaka, A., Williams, S.C., Richardson, J.A., Tsujino, N., Garry, M.G., Lerner, M.R., King, D.S., O'Dowd, B.F., Sakurai, T., and Yanagisawa, M. (2003). Characterization of a family of endogenous neuropeptide ligands for the G protein-coupled receptors GPR7 and GPR8. Proc Natl Acad Sci U S A 100, 6251–6256.

Tatemoto, K., Hosoya, M., Habata, Y., Fujii, R., Kakegawa, T., Zou, M.X., Kawamata, Y., Fukusumi, S., Hinuma, S., Kitada, C., Kurokawa, T., Onda, H., and Fujino, M. (1998). Isolation and characterization of a novel endogenous peptide ligand for the human APJ receptor. Biochem Biophys Res Commun 251, 471–476.

Thannickal, T.C., Moore, R.Y., Nienhuis, R., Ramanathan, L., Gulyani, S., Aldrich, M., Cornford, M., and Siegel, J.M. (2000). Reduced number of hypocretin neurons in human narcolepsy. Neuron 27, 469–474.

Thorpe, A.J., Mullett, M.A., Wang, C., and Kotz, C.M. (2003). Peptides that regulate food intake: regional, metabolic, and circadian specificity of lateral hypothalamic orexin A feeding stimulation. Am J Physiol Regul Integr Comp Physiol 284, R1409–1417.

Wang, J., Osaka, T., and Inoue, S. (2001). Energy expenditure by intracerebroventricular administration of orexin to anesthetized rats. Neurosci Lett 315, 49–52.

Wang, J., Osaka, T., and Inoue, S. (2003). Orexin-A-sensitive site for energy expenditure localized in the arcuate nucleus of the hypothalamus. Brain Res 971, 128–134.

Willie, J.T., Chemelli, R.M., Sinton, C.M., Tokita, S., Williams, S.C., Kisanuki, Y.Y., Marcus, J.N., Lee, C., Elmquist, J.K., Kohlmeier, K.A., Leonard, C.S., Richardson, J.A., Hammer, R.E., and Yanagisawa, M. (2003). Distinct narcolepsy syndromes in Orexin receptor-2 and Orexin null mice: molecular genetic dissection of Non-REM and REM sleep regulatory processes. Neuron 38, 715–730.

Willie, J.T., Chemelli, R.M., Sinton, C.M., and Yanagisawa, M. (2001). To eat or to sleep? Orexin in the regulation of feeding and wakefulness. Annu Rev Neurosci 24, 429–458.

Xu, Y.L., Reinscheid, R.K., Huitron-Resendiz, S., Clark, S.D., Wang, Z., Lin, S.H., Brucher, F.A., Zeng, J., Ly, N.K., Henriksen, S.J., de Lecea, L., and Civelli, O. (2004). Neuropeptide S: a neuropeptide promoting arousal and anxiolytic-like effects. Neuron 43, 487–497.

Yamamoto, Y., Ueta, Y., Date, Y., Nakazato, M., Hara, Y., Serino, R., Nomura, M., Shibuya, I., Matsukura, S., and Yamashita, H. (1999). Down regulation of the prepro-orexin gene expression in genetically obese mice. Brain Res Mol Brain Res 65, 14–22.

Yamanaka, A., Beuckmann, C.T., Willie, J.T., Hara, J., Tsujino, N., Mieda, M., Tominaga, M., Yagami, K., Sugiyama, F., Goto, K., Yanagisawa, M., and

Sakurai, T. (2003). Hypothalamic orexin neurons regulate arousal according to energy balance in mice. Neuron *38*, 701–713.

Yamanaka, A., Sakurai, T., Katsumoto, T., Yanagisawa, M., and Goto, K. (1999). Chronic intracerebroventricular administration of orexin-A to rats increases food intake in daytime, but has no effect on body weight. Brain Res *849*, 248–252.

Neuronal Mechanisms of Feeding Regulation by Peptides

Masamitsu Nakazato and Hiroaki Ueno

Neurology, Respirology, Endocrinology and Metabolism, Miyazaki Medical College, University of Miyazaki, 5200, Kiyotake, Miyazaki 889-1692, Japan
<e-mail> nakazato@med. miyazaki-u. ac. jp

Summary. Understanding the feeding regulatory mechanisms that occur in upstream of obesity is important for the effective treatment of life-style related diseases. Feeding is regulated by complex interactions between orexigenic substances and anorectic substances, which are produced in the central nervous system and peripheral organs. Information about satiety or hunger is transmitted humorally and neurogenically to the brain, where it is integrated with various neural processes in the hypothalamus, such as learning, memory, cognition, and motion perception, and affects the feeding enhancement system or feeding inhibitory system. For feeding regulation, information is obtained not only from the central nervous system, including the hypothalamus and the cerebral limbic system, but also from peripheral organs, including the gastrointestinal tract, liver and adipose tissue. Afferent information from the gastrointestinal tract, liver, and adipose tissue concerning feeding and energy homeostasis is transmitted to the brain stem and hypothalamus through the vagus nerve and the circulatory system. Interestingly, the vagus afferent nerve expresses various receptors for substances that regulate feeding and is controlled by multiple factors.

Keywords. cholecystokinin, ghrelin, hypothalamus, the nucleus of the solitary tract, vagus nerve

1 Introduction

A connection between peripheral organs, especially the gastrointestinal (GI) tract, and hunger was first considered at least 160 years ago. In 1912, Cannon and Washburn demonstrated that hunger pangs closely corresponded with gastric contractions in humans. Carlson reported in 1916 that the absence of gastric contents induced hunger, whereas the presence of gastric contents induced satiety. In 1973, cholecystokinin (CCK), one of the peptides produced in the GI tract, was the first hormone that was shown to decrease meal size in rats (Gibbs et al. 1973); thereafter, a number of other GI hormones that regulate appetite have been identified (Table 1). We now know that information concerning feeding behavior and energy metabolism moves from peripheral organs to the central nervous system not only through the bloodstream but also through the vagus nerve.

2 Sensors in the GI tract transmit satiety and hunger signals to the brain

2-1 Mechanosensors

When a rat's stomach is distended using a balloon, the expression of c-Fos, a marker of neuronal activation, increases in the nucleus of the solitary tract (NTS) and the dorsal motor nucleus of the medulla (Willing and Berthoud 1997). This elevation of c-Fos expression in the NTS is blocked by vagotomy and perivagal capsaicin treatment, suggesting that the vagal afferent pathway mediates this response. Pretreatment with a serotonin (5-hydroxytryptamine; 5-HT) 5-HT3 receptor antagonist also blocks gastric distension-induced c-Fos expression in the NTS, suggesting that 5-HT released from the enterochromaffin cells of the stomach activates 5-HT3 receptors located in the peripheral vagal afferent nerve terminals, thereby inducing c-Fos expression. In addition, 20~30% of the NTS neurons that produce pre-proglucagon, a precursor of glucagon-like peptide-1 (GLP-1), but none or only a small population of the catecholaminergic neurons in the NTS are activated after gastric distension. GLP-1-containing neurons in the NTS project to several hypothalamic areas involved in the regulation of food intake. Mechanosensory GI tract stimuli are relayed preferentially in the GLP-1-containing neurons of the NTS through the vagal afferent nerves, after which the information is conveyed to the hypothalamus. The GLP-1-containing neurons, however, are only located in the NTS in the central nervous system, suggesting that the central anorectic actions of

GLP-1 are triggered by the activation of mechanosensory GI tract afferents.

Leptin, a 16-kDa protein produced by white adipose tissue, decreases adiposity and body weight by reducing appetite and food intake. Leptin receptors are found in rodent NTS neurons, and low-dose leptin injections directly into the NTS reduce food intake. Approximately 40% of the leptin-sensitive neurons in the rodent NTS are activated by mechanical

Table 1. Feeding regulatory substances

	Anorectic substances		Orexigenic substances	
	Substance	Main expression area	Substance	Main expression area
Central nervous system	α-MSH	ARC	AgRP	ARC
	CART	ARC	Anandamide	Basal nuclei, Limbic system
	CRH	PVN	Galanin	ARC, PVN
	Histamin	Tuberomammilary nucleus	GALP	ARC
	Neuromedin U	ARC	MCH	LHA
	Noradrenaline (α1,β)	Caeruleun nucleus, NTS	Noradrenaline (α2)	Caeruleun nucleus, NTS
	NPB	Midbrain, Hippocampus	NPY	ARC, PVN
	NPW	PVN, Supraoptic nucleus	Orexin	LHA
	POMC	ARC		
	PrRP	Ventromedial nucleus, NTS		
	Serotonin	Raphe nuclei		
	Urocortin	Midbrain, Supraoptic nucleus		
	Urocortin II	ARC, PVN		
	Urocortin III	Ventromedial nucleus, PVN		
Peripheral organ	Cholecystokinin	Upper intestine	Ghrelin	Stomach
	GLP-1	Lower intestine		
	Insulin	Pancreatic β cell		
	Leptin	Adipose cell		
	Oxyntomodulin	Lower intestine		
	PYY_{3-36}	Colon, Rectum		

α-MSH, α melanocyte stimulating hormone; ARC, Arecuate nucleus; AgRP, Agouti gene-related protein; CART, cocain- and amphetamine-regulated transcript; CRH, corticotropin releasing hormone; GALP, Galanin like peptide; GLP-1, glucagon like peptide-1; LH, Lateral hypothalamic area; MCH, Melanocortin concentrating hormone; NPB, Neuropeptide B; NPW, Neuropeptide W; NPY, neuropeptide Y; NTS, Nucleus of solitary tract; POMC, proopiomelanocortin; PrRP, Prolactin related protein; PVN, Paraventricular nucleus.

distension of the stomach. In addition, the combination of a low-dose injection of leptin into the brainstem and small mechanical gastric distension, each of which has no effect when applied separately, reduces food intake in rats (Grill et al. 2002). These results suggest that leptin reduces food intake not only by acting in the hypothalamus, but also in the NTS through an interaction with signals associated with gastric distension. Because pre-proglucagon producing neurons in the NTS coexpress leptin receptors, leptin also contributes to GLP-1-mediated signaling.

2-2 Chemoreceptors

The mucosal terminals are mostly chemosensory and are regulated by various substances, including glucose, amino acids, fatty acids with different chain lengths, and gut hormones such as CCK, peptide YY $(PYY)_{3-36}$, somatostatin, and GLP-1. Although gastric distension activates the pre-proglucagon producing neurons of the NTS, information from chemoreceptors is relayed preferentially to the catecholaminergic neurons of the NTS via vagal afferent nerves. The catecholaminergic neurons of the NTS, which express leptin receptors, project to the hypothalamus, central amygdaloid nucleus, and the bed nucleus of the stria terminalis in rats. The feeding regulatory functions of the gut hormones are described in a later section.

3 Gut hormone-mediated feeding regulation via the vagus nerve

3-1 Ghrelin

Ghrelin is mainly produced in endocrine cells (Gh cells) in the stomach and is secreted into the blood (Kojima et al. 1999). When ghrelin is administered to rats peripherally, dose-dependent increases in feeding occur starting at a concentration of 1.5 nmol, whereas the administration of ghrelin-neutralizing antibodies decreases food intake. This action occurs independently of the stimulatory effect of ghrelin on growth hormone (GH) secretion (Nakazato et al. 2001). In addition, ghrelin suppresses sympathetic nerve activity and concurrently promotes parasympathetic nerve activity in order to reduce energy consumption. Ghrelin receptors are expressed in nuclei in the hypothalamus, hippocampus, nigra in the midbrain, and other areas. Both peripheral and central administration of ghrelin-stimulates food intake. Currently, ghrelin is the only orexigenic peptide that has been detected in peripheral organs.

Ghrelin transmits information regarding feeding and GH secretion to the hypothalamus via the vagal afferent pathway. When ghrelin is intravenously administered to rats that have undergone a vagotomy or capsaicin treatment, the enhancement of feeding and GH secretion are not observed. Intravenous administration of ghrelin does not induce expression of c-Fos in neurons in the arcuate nucleus of the hypothalamus that contain NPY and GH-releasing hormone in these rat models. Ghrelin receptors are produced in the cell bodies of the nodose ganglion and are axonally transported to the vagal afferent fiber terminals. At these locations, ghrelin binds to the ghrelin receptor, and inhibits the electrical activity of the gastric branches of the vagal afferent nerves (Date et al. 2002) (Fig. 1).

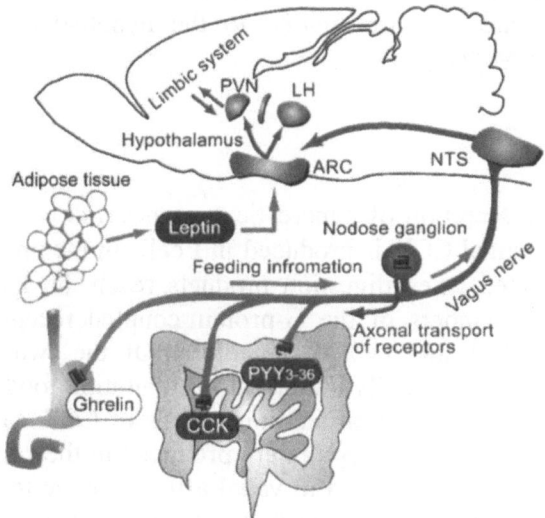

Fig. 1. The signal transduction pathways from the peripheral organs to the hypothalamus mediated by feeding regulatory substances. Receptors for ghrelin, cholecystokinin (CCK), and peptide YY (PYY)$_{3-36}$ are produced in the nodose ganglion and transported to the gastric and intestinal mucosa via axonal transport through the vagal afferent nerve. After binding to each receptor, ghrelin, CCK, and PYY$_{3-36}$ change the electrical activity of the vagal afferent fibers, and these electrical signals are transmitted to the NTS, where synapses transmit the signals to secondary neurons. Secondary neurons project to the hypothalamic arcuate nucleus. Information about feeding from peripheral organs and the central nervous system, including the limbic system, is integrated in the hypothalamus, before decisions regarding feeding actions are made. ARC, arcuate nucleus; LH, lateral hypothalamus; NTS, nucleus of the solitary tract; PVN, paraventricular nucleus

Ghrelin-mediated signals are transmitted to the hypothalamus through noradrenaline neurons that act as secondary neurons, following transmission to the NTS. Peripheral administration of ghrelin increases noradrenaline release in the arcuate nucleus. On the other hand, intracerebroventricular pretreatment with a noradrenalin receptor α1 or β2 blocker and subsequent peripheral administration of ghrelin does not stimulate food intake. Furthermore, when the noradrenaline neurons of the NTS projecting to the arcuate nucleus are destroyed using saporin toxin, the effects of peripherally administered ghrelin disappear. Even when ghrelin is peripherally administered to rats that had undergone a midbrain transection by blocking the nerve fibers from the medulla to the hypothalamus, feeding enhancement does not occur and noradrenaline release in the arcuate nucleus is not observed (Date et al. 2006). These results suggest that ghrelin-mediated signals from the stomach reach the NTS via the vagus afferent nerves and are then transmitted to the hypothalamus through the noradrenaline nerves (Fig. 1).

3-2 CCK

CCK affects the secretion of pancreatic enzymes and contractions of the gallbladder. Intestinal CCK is produced in I cells in the upper small intestine and is secreted when digestion products reach the intestinal lumen. CCK receptors, members of the G-protein-coupled receptor family, are classified as CCK-1 and CCK-2 receptors. Of the two receptors, the CCK-1 receptor is functionally important for inhibiting food intake.

Both peripheral and central administrations of CCK cause short-term feeding inhibition. CCK-1 receptors are produced in the cell bodies of the nodose ganglion and transmitted to vagal afferent nerve terminals via axonal transport. Peripheral administration of CCK in rats increases electrical activity in vagal afferent fibers. CCK-mediated signals regarding feeding inhibition reach the NTS, and are transmitted to the hypothalamus through secondary neurons (Fig. 1). In fact, feeding inhibition due to the peripheral administration of CCK is not observed in vagotomized rats (Smith et al. 1981).

Recent studies have suggested that several other neurohumoral factors may modulate the effects of CCK on vagal afferent nerve discharge and food intake. Some vagal afferent neurons coexpress leptin receptors and CCK-1 receptors, and synergistic interactions have been observed between CCK and leptin for both the inhibition of food intake and vagal afferent nerve discharge. Conversely, some CCK-1 receptor-expressing neurons also express receptors for the orexigenic peptide orexin A, which inhibits the stimulatory effect of CCK on vagal afferent nerve discharge. The cannabinoid receptor CB1 and its endogenous ligands are associated with the stimulation of appetite via central and peripheral mechanisms. Cannabinoid CB1 receptors, the orexigenic peptide melanin-concentrating hormone (MCH), and its receptor MCH-1R are also expressed in the nodose ganglion as well as the central nervous system. The expression of CB1 receptor, MCH, and MCH-1R is stimulated by fasting and downregulated by refeeding in a CCK-dependent manner. Moreover, approximately 70% of ghrelin receptor-containing neurons coexpress CCK-1 receptor in the rat nodose ganglion. CCK inhibits the ghrelin-induced stimulatory effect on food intake and c-Fos expression in the arcuate nucleus. Furthermore, CCK and ghrelin mutually modulate their effects on vagal afferent activity (Date et al. 2005).

3-3 PYY

PYY consists of 36 amino-acid residues and belongs to the NPY peptide family, which includes NPY and pancreatic polypeptide. PYY produced in L cells primarily in the ileum is secreted mainly as PYY_{3-36} after eating. Peripheral administration of PYY_{3-36} reduces food intake in both human and rodents. In vagotomized rats, however, the inhibition of food intake or expression of c-Fos in the arcuate nucleus were not observed after peripheral administration of PYY_{3-36}. PYY_{3-36} binds with high affinity to one of the members of the NPY receptor family, Y2 receptor (Y2-R). PYY_{3-36} is thought to reduce food intake by binding to autoinhibitory presynaptic Y2-R on NPY neurons in the arcuate nucleus of the hypothalamu. PYY_{3-36} has no effects on food intake in Y2-R-deficient mice, suggesting that its effects on appetite are mediated by Y2-R-regulated pathways. Y2-R is also produced in the cell bodies of the nodose ganglion and is axonally transported to vagal afferent terminals. Peripheral administration of PYY_{3-36} has been observed to increase electrical activity in vagal afferent fibers. The vagal afferent pathway is thought to convey satiety signals mediated by PYY_{3-36} to the hypothalamus (Koda et al. 2005) (Fig. 1).

4 Conclusion and perspective

The vagus nerve plays important roles in the transmission of regulatory information about feeding behavior and energy homeostasis between peripheral organs and the central nervous system. The development and practical application of antiobesity drugs based on the mechanisms of feeding regulation are expected to become a potent strategy for the treatment of modern lifestyle-related diseases.

References

Date Y, Murakami N, Toshinai K, Matsukura S, Niijima A, Matsuo H, Kangawa K, Nakazato M (2002) The role of the gastric afferent vagal nerve in ghrelin-induced feeding and growth hormone secretion in rats. Gastroenterology 123: 1120–1128

Date Y, Toshinai K, Koda S, Miyazato M, Shimbara T, Tsuruta T, Niijima A, Kangawa K, Nakazato M (2005) Peripheral interaction of ghrelin with cholecystokinin on feeding regulation. Endocrinology 146: 3518–3525

Date Y, Shimbara T, Koda S, Toshinai K, Ida T, Murakami N, Murakami N, Miyazato M, Kokame K, Ishizuka Y, Ishida Y, Kageyama H, Shioda S, Kangawa K, Nakazato M (2006) Peripheral ghrelin transmits orexigenic signals through the noradrenergic pathway from the hindbrain to the hypothalamus. Cell Metab 4: 323–331

Gibbs J, Young RC, Smith GP (1973) Cholecystokinin decreases food intake in rats. J Comp Physiol Psychol 84: 488–495

Grill HJ, Schwartz MW, Kaplan JM, Foxhall JS, Breininger J, Baskin DG (2002) Evidence that the caudal brainstem is a target for the inhibitory effect of leptin on food intake. Endocrinology 143: 239–246

Koda S, Date Y, Murakami N, Shimbara T, Hanada T, Toshinai K, Niijima A, Furuya M, Inomata N, Osuye K, Nakazato M (2005) The role of the vagal nerve in peripheral PYY3-36-induced feeding reduction in rats. Endocrinology 146: 2369–2375

Kojima M, Hosoda H, Date Y, Nakazato M, Matsuo H, Kangawa K (1999) Ghrelin is a growth-hormone-releasing acylated peptide from stomach. Nature 402: 656–660

Nakazato M, Murakami N, Date Y, Kojima M, Matsuo H, Kangawa K, Matsukura S (2001) A role for ghrelin in the central regulation of feeding. Nature 409: 194–198

Smith GP, Jerome C, Cushin BJ, Eterno R, Simansky KJ (1981) Abdominal vagotomy blocks the satiety effect of cholecystokinin in the rat. Science 213: 1036–1037

Willing AE, Berthoud HR (1997) Gastric distension-induced c-fos expression in catecholaminergic neurons of rat dorsal vagal complex. Am J Physiol 272: R59–R67

Distribution and Localization of Galanin-like peptide (GALP) in Brain

Fumiko Takenoya[1, 2], Haruaki Kageyama[1], Seiji Shioda[1]

[1]Department of Anatomy I, Showa University School of Medicine, 1-5-8 Hatanodai, Shinagawa-ku, Tokyo 142-8555, Japan
<e-mail> haruaki@med.showa-u.ac.jp, shioda@med.showa-u.ac.jp
[2]Department of Physical Education, Hoshi University School of Pharmacy and Pharmaceutical Science, 2-4-41 Ebara, Shinagawa, Tokyo 142-8501, Japan
<e-mail> kuki@hoshi.ac.jp

Summary. Galanin-like peptide (GALP) is a hypothalamic neuropeptide which was isolated as an endogenous ligand for galanin receptors. GALP-containing neurons are shown to be widely distributed in brain. However, GALP and galanin are assumed to act through a different receptor pathway to exert its each effects on feeding regulation. Therefore, it is suggested that GALP has its own specific receptor but its specific receptor has not yet been identified. Recently physiological studies have revealed that GALP has many functions including energy metabolism, thermoregulation and reproduction as well as feeding regulation. We and others have reported that GALP-containing neurons make different input and output from a variety of neurons which contain several kind of neurotransmitters and/or neuromodulators to maintain homeostasis in feeding and energy metabolism. In this chapter, we will review the neuron network between GALP-containing neurons and other orexigenic and/or anorexic neuropeptide-containing neurons in the hypothalamus based on our morphological observation.

Key words. galanin-like peptide (GALP), hypothalamus, feeding, neuron network

1. Introduction

In 1983, galanin was discovered as a 29-30 amino acid peptide which was widely distributed throughout the central nervous system (CNS) and was classified as an orexigenic peptide (Tatemoto et al., 1983). Its receptors belong to a member of G-protein-coupled receptor (GPCR) family and have three subtypes, GalR1 (Habert-Ortoli et al., 1994; Parker et al., 1995), GalR2 (Howard et al., 1997) and GalR3 (Branchek et al., 2000; Wang et al., 1997).

In 1999, GALP was discovered as a 60 amino acid peptide, isolated from the porcine hypothalamus on the basis of its ability to bind and activate galanin receptors *in vitro* (Ohtaki et al., 1999). This is the reason why this peptide was named as GALP. GALP binds to galanin receptor type 1 (GALR1) with a lower affinity than galanin, but it binds to galanin receptor type 2 (GALR2) with a higher affinity than galanin (Gundlach, 2002). The reason for the difference of the action of GALP and galanin, when they were centrally administered, is assumed to produce different pattern of c-Fos in brain (Fraley et al., 2003; Lawrence et al., 2002), and several effects of physiological systems such as feeding (Krasnow et al., 2003; Lawrence et al., 2002), luteinizing hormone (LH) secretion, and sexual behavior. These results suggest that GALP has its own specific receptor. A physiological study has reported that intracerebroventricular (i.c.v.) infusion of GALP has an orexigenic effect during a short time (Matsumoto et al., 2002). But 24 hours after GALP i.c.v. infusion, cumulative food intake and body weight are reduced in rat (Krasnow et al., 2003; Lawrence et al., 2002). In contrast, in mouse, GALP administration decreases food intake during 2 hours, and suppresses both food intake and body weight for approximately 24 hours (Krasnow et al., 2003). In *ob/ob* mice, sustained decrease in food intake and body weight for a longer time of i.c.v. infusion (Hansen et al., 2003). For this reason, it is expected that the feeding function of GALP is complex and this peptide has a multiple function on feeding regulation. Recently, it is demonstrated that GALP is involved in reproduction in addition to feeding regulation (Matsumoto et al., 2001; Seth et al., 2004). Therefore, GALP is a peptide which links both feeding and reproductive function, and it has neuronal interactions between other peptides of hormones which regulate reproductive functions as well as feeding. This review will describe the distribution and neuronal networks of GALP in brain especially in the hypothalamus.

2. Distribution of GALP-containing neurons in brain

GALP mRNA is distributed in the periventricular region of the arcuate nucleus (ARC) (Jureus et al., 2000, Kerr et al., 2000, Larm and Gundlach, 2000) and in the median eminence (Jureus et al., 2000, Kerr et al., 2000) and pituitary gland (Kerr et al., 2000) of rat by using *in situ* hybridization. Takatsu et al. (Takatsu et al., 2001) have reported, by using immunohistochemistry with a monoclonal antibody against the N-terminal sequence of GALP, that GALP-immunoreactive neuronal cell bodies are shown to be present in the ARC particularly densely concentrated in its medial to posterior portion and in the posterior pituitary (Fujiwara et al., 2002; Takatsu et al., 2001). Subsequently, we have shown that GALP-immunoreactive cell bodies are distributed mainly in the ARC, which the number of GALP-positive cell bodies are abundant in the posterior part of the ARC rather than in its rostral part (Takenoya et al., 2002) (Fig1-A). On the other hand, GALP-immunoreactive nerve fibers originating from the rostral to caudal part of the ARC are found in the ARC as well as in the paraventricular nucleus (PVN), stria terminalis (BST), medial preoptic area (MPA) and lateral septal nucleus (LSV). In addition, we have reported that GALP-immunoreactive fibers are also found in the lateral hypothalamus (LH) near the fornix (Takenoya et al., 2005). Thus, GALP-positive fibers widely project to other parts of the hypothalamus but they also project to the extra-hypothalamus including the thalamus and limbic system. In the ARC, GALP-positive axon terminals are shown to make synaptic inputs to other unknown neurons by use of electron microscopy (EM) (Guan et al., 2005), suggesting that GALP may function as a neurotransmitter and/or neuromodulator in the CNS.

3. c-Fos expression in brain after treatment with GALP

Central administration of GALP and galanin into brain has showed differential patterns of c-Fos immunoreactivity. Galanin is shown to induce significantly a large number of c-Fos-positive neurons in the hypothalamic PVN as compared with GALP, however, GALP is shown to induce large number of c-Fos-positive neurons in the horizontal limb of the diagonal band of Broca, caudal preoptic area, ARC and median eminence (Fraley et al., 2003). Moreover, c-Fos-like immunoreactivity is intensely expressed in GALP-positive neurons in the hypothalamus after foot shock stress (Onaka et al., 2005). Furthermore, Lawrence et al. (Lawrence et al., 2002) have reported that i.c.v. infusion of GALP intensely stimulates c-Fos-like immunoreactivity in astrocytes but not in microglial cells in the hypothalamic region.

4. Neuronal interaction between GALP and other peptides in brain

To understand the neuron network of GALP in brain, several groups have tried to clarify the neuron network of GALP in rat brain by using immunohistochemical methods. Neurons in the ARC region are known to

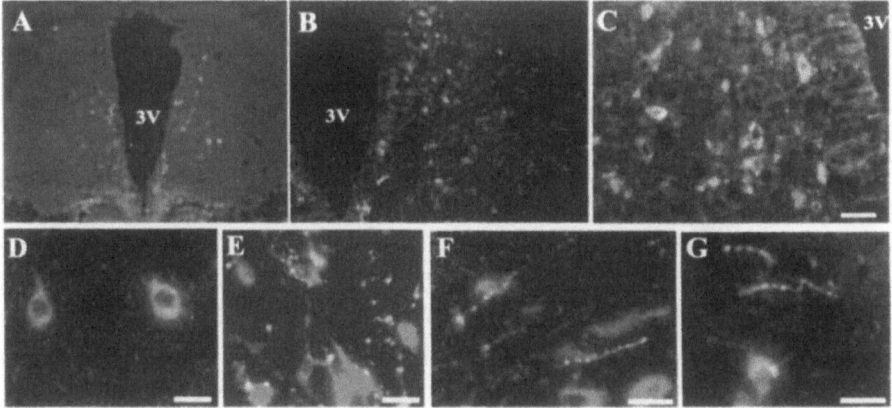

Fig. 1. Immunofluorescence photomicrographs of the arcuate nucleus (ARC) after a direct double-labeling method combining antiserum against galanin-like peptide (GALP) with antiserum against several kind of neuropeptides. A; Immunoreactivity for GALP (red) is localized in the ARC. B; Immunoreactivity for GALP (red) and α-melanocyte-stimulating hormone (MSH) (green). α-MSH-immunoreactive neurons are co-localized with GALP. C; Immunoreactivity for GALP (green) and orexin type 1 receptor (red). GALP immunoreactive neurons in the ARC express orexin type 1 receptor. D; Immunoreactivity for GALP (red) and neuropeptide Y (NPY) (green). NPY-immunoreactive fibers are close apposition with GALP-immunoreactive neurons in the ARC. E; Immunoreactivity for GALP (green) and orexin (red). Orexin-immunoreactive fibers are closely apposed to GALP-like immunoreactive neurons in the ARC. F; Immunoreactivity for GALP (red) and orexin (green). GALP-immunoreactive fibers are in direct contact with orexin-immunoreactive neurons in the LH. G; Immunoreactivity for GALP (red) and melanin-concentrating hormone (MCH) (green). GALP-immunoreactive fibers make direct contact on MCH-immunoreactive neurons in the LH. 3V; third ventricle, Bars; 20 μ m.

express a high concentration of immunoreactivity for leptin receptor (leptin-R). Using a double immunostaining method, GALP and leptin-R-like immunoreactivity are shown to co-localize in the same neurons in the ARC, approximately 85% of the GALP-immunoreactive neurons express leptin-R-like immunoreactivity (Takatsu et al., 2001). Several studies have reported the effect of leptin and/or insulin on GALP-containing neurons. In the ARC, gene expression of GALP mRNA is fist reduced but recovered in the ARC after treatment of leptin (Jureus et al., 2000). Moreover, *ob/ob* mice express a reduction of GALP mRNA in the hypothalamus (Jureus et al., 2001; Kumano et al., 2003). Fraley et al. (Fraley et al., 2004) have shown that streptozotocin-induced diabetes is associated with a significant reduction in gene expression of GALP mRNA, which is reversed by treatment with either insulin. Moreover, in the fasted rats, GALP mRNA appears lower level during 48 hours but the central treatment of insulin reverses the level of GALP mRNA (Fraley et al., 2004).

In addition, our observations indicate that GALP-like immunoreactivity is co-expressed with α-melanocyte-stimulating hormone (α-MSH)-like immunoreactivity, which is derived from pro-opiomelanocortin (POMC) (Fig1. B) (Takenoya et al., 2002). We have also showed that about 10% of GALP-immunoreactive cell bodies are overlapped with POMC-immunoreactive cell bodies which end product is α-MCH that is a strong inhibitory effect on feeding. These double-labeling cell bodies are distributed intensely in the caudal part of the ARC. We have also found that approximately 10% of GALP-positive neurons express orexin-1 receptor-like immunoreactivity (Fig1. C) (Takenoya et al., 2003) which is an orexigenic action from orexin-A (Rodgers et al., 2002). Many GALP-positive neurons, expressing orexin-1 receptor-like immunoreactivity, are distributed densely in the rostral portion of the ARC as compared to its medial portion. However, co-expression of GALP with both α-MCH and orexin-1 receptor is very low percentage in the ARC. In addition, Cunningham et al. (Cunningham et al., 2004) have reported that GALP-positive neurons co-express mRNA for NPY Y1 receptor (approximately 40%) and for serotonin 5-HT2C receptor (approximately 25%) by using a double-labeling *in situ* hybridization with immunohistochemistry on the tissue sections of macaque brain. We are now trying to determine the afferent and efferent neurons of GALP in mouse and rat brain.

As to the afferent inputs to GALP-containing neurons in rat brrain, we have found that neuropeptide Y (NPY)- and orexin-containing axon terminals are in close apposition with GALP-containing neurons in the ARC (Takenoya et al., 2003; Takenoya et al., 2002) (Fig1-D, E). As to the efferent output from GALP-containing neurons in the ARC, GALP-like immunoreactive nerve fibers are directly contacted with orexin- and/or MCH-like immunoreactive neurons in the LH (Takenoya et al., 2005) (Fig1-F, G). Kageyama et al. (Kageyama et al., 2006) have tried to determine whether GALP regulates feeding behavior via orexin neurons or not, GALP i.c.v. infusion with or without anti-orexin immunoglobulin (IgG) pretreatment is undertaken. The anti-orexin IgGs markedly inhibit GALP-induced hyperphagia, suggesting that orexin-containing neurons in the LH are targeted by GALP. Additionally, at the EM level, GALP-immunoreactive axon terminals are shown to make synapses on orexin-immunoreactive cell bodies and dendritic processes in the LH, suggesting that GALP may regulate orexin neurons (Kageyama, 2006). Another reports appear to reveal the function of GALP on feeding regulation, i.c.v. GALP infusion induces a potent short-term stimulation of food intake mainly via activation of NPY-containing neurons in the dorsomedial nucleus (DMH) (Kuramochi et al., 2006). These findings strongly suggest that GALP-containing neurons make direct contact with orexigenic and/or anorexic neurons and GALP-containing neurons may participate in regulation of feeding behavior in harmony with other feeding-regulating neurons in the hypothalamus.

It has been reported that administration of GALP in the PVH decreases the level of plasma thyroid-stimulating hormone (TSH) (Seth et al., 2003). Moreover, small number of GALP-immunoreactive axon terminals are associated with thyroid-stimulating hormone-releasing hormone (TRH)-containing neurons in the medial parvocellular region of the PVH which is associated with regulation of food intake (Wittmann et al., 2004). These results suggest that TRH-containing neurons make it unlikely that GALP similarly exerts direct regulatory effects on hypophysiotropic, and GALP may influence the hypothalamo–pituitary–thyroid axis via an indirect mechanism to suppress circulating TSH levels.

Moreover, GALP is shown to increase plasma LH through luteinizing hormone-releasing hormone (LHRH) (Matsumoto et al., 2001). Recently, using transgenic rat, we have tired to study the neuronal interaction between GALP- and LHRH-containing neurons. As a result, we have found that GALP-positive axon terminals lie in close apposition with LHRH-containing cell bodies and processes in the medial preoptic area (MPA) (Fig.2, A, B) and GALP-positive axon terminals are in direct contact with LHRH-immunoreactive nerve processes in the bed nucleus of the stria terminalis (BST) (Fig.2, C) (Takenoya et al., 2006). At the EM level, the GALP-positive axon terminals are found to make axo-somatic (90.2%) and axo-dendritic (9.8%) synaptic contacts with the LHRH-neurons (Takenoya et al., 2006). These finding indicate that GALP-containing neurons make direct inputs to LHRH-containing neurons in brain, and GALP may play a crucial role in the regulation of LH secretion via LHRH.

In conclusion, these our and other results suggest that GALP has a complicated neuron network in brain, and its function is feeding behavior and energy homeostasis as well as reproduction in harmony with other neurons in the hypothalamus (Fig.2).

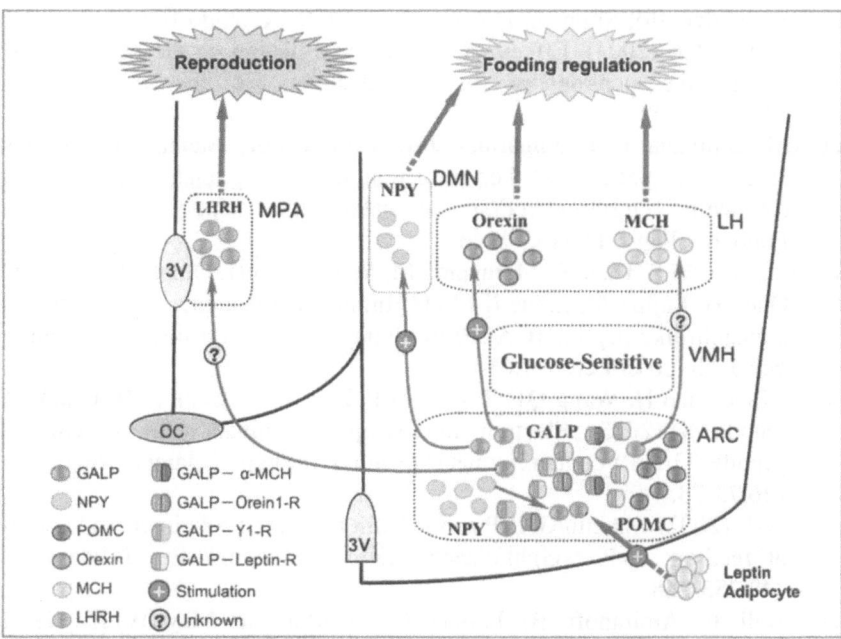

Fig. 2. Schematic illustration based on the findings and reproduction of morphological and physiological studies in the hypothalamus by neurons containing peptides. The plus or minus indicates stimulatory or inhibitory effects, respectively. Question marks indicate still unsolved issues, whether stimulatory or inhibitory. NPY; neuropeptide Y, POMC; pro-opiomelanocortin, LHRH; luteinizing hormone-releasing hormonem, MCH: melanin-concentrating hormone, α -MSH; alpha-melanocyte-stimulating hormone, Orexin1-R; orexin type 1 receptor, Leptin-R; leptin receptor, 3V; third ventricle.

Acknowledgement

The authors thank Dr. Tetsuya Ohtaki in Takeda Chemical Industrial Company and Ms. Sachi Kato for their help to accomplish this study. This study was supported in part by a grant from the Ministry of Education, Culture, Sports, Science and Technology of Japan

References

Branchek TA, Smith KE, Gerald C, Walker MW (2000) Galanin receptor subtypes. Trends Pharmacol Sci 21:109-117.

Cunningham MJ, Shahab M, Grove KL, Scarlett JM, Plant TM, Cameron JL, Smith MS, Clifton DK, Steiner RA (2004) Galanin-like peptide as a possible link between metabolism and reproduction in the macaque. J Clin Endocrinol Metab 89:1760-1766.

Fraley GS, Scarlett JM, Shimada I, Teklemichael DN, Acohido BV, Clifton DK, Steiner RA (2004). Effects of diabetes and insulin on the expression of galanin-like peptide in the hypothalamus of the rat. Diabetes 53:1237-1242.

Fraley GS, Shimada I, Baumgartner JW, Clifton DK, Steiner, RA (2003) Differential patterns of Fos induction in the hypothalamus of the rat following central injections of galanin-like peptide and galanin. Endocrinology 144:1143-1146.

Fujiwara K, Adachi S, Usui K, Maruyama M, Matsumoto H, Ohtaki T, Kitada C, Onda H, Fujino M, Inoue K (2002) Immunocytochemical localization of a galanin-like peptide (GALP) in pituicytes of the rat posterior pituitary gland. Neurosci Lett 317:65-68.

Guan JL, Kageyama H, Wang QP, Takenoya F, Kita T, Matsumoto H, Ohtaki T, Shioda S (2005) Electron microscopy examination of galanin-like peptide (GALP)-containing neurons in the rat hypothalamus. Regul Pept 126:73-78.

Gundlach AL (2002) Galanin/GALP and galanin receptors: role in central control of feeding, body weight/obesity and reproduction? Eur J Pharmacol 440:255-268.

Habert-Ortoli E, Amiranoff B, Loquet I, Laburthe M, Mayaux JF (1994) Molecular cloning of a functional human galanin receptor. Proc Natl Acad Sci U S A 91:9780-9783.

Hansen KR, Krasnow SM, Nolan MA, Fraley GS, Baumgartner JW, Clifton DK, Steiner RA (2003) Activation of the sympathetic nervous system by galanin-like peptide--a possible link between leptin and metabolism. Endocrinology 144:4709-4717.

Howard AD, Tan C, Shiao LL, Palyha OC, McKee KK, Weinberg DH, Feighner, SD, Cascieri MA, Smith RG, Van Der Ploeg LH, Sullivan KA (1997) Molecular cloning and characterization of a new receptor for galanin. FEBS Lett 405:285-290.

Jureus A, Cunningham MJ, Li D, Johnson LL, Krasnow SM, Teklemichael DN, Clifton DK, Steiner RA (2001) Distribution and regulation of galanin-like peptide (GALP) in the hypothalamus of the mouse. Endocrinology 142:5140-5144.

Jureus A, Cunningham MJ, McClain ME, Clifton DK, Steiner RA (2000) Galanin-like peptide (GALP) is a target for regulation by leptin in the hypothalamus of the rat. Endocrinology 141:2703-2706.

Kageyama H (2006) Galanin-Like Peptide promotes feeding behaviour via activation of orexinergic neurones in the rat lateral hypothalamus. J Neuroendocrinol 18:33-41.

Kerr NC, Holmes FE, Wynick D (2000). Galanin-like peptide (GALP) is expressed in rat hypothalamus and pituitary, but not in DRG. Neuroreport 11:3909-3913.

Krasnow SM, Fraley GS, Schuh SM, Baumgartner JW, Clifton DK, Steiner RA (2003) A role for galanin-like peptide in the integration of feeding, body weight regulation, and reproduction in the mouse. Endocrinology 144:813-822.

Kumano S, Matsumoto H, Takatsu Y, Noguchi J, Kitada C, Ohtaki T (2003) Changes in hypothalamic expression levels of galanin-like peptide in rat and mouse models support that it is a leptin-target peptide. Endocrinology 144:2634-2643.

Kuramochi M, Onaka T, Kohno D, Kato S, Yada, T., 2006. Galanin-like peptide stimulates food intake via activation of neuropeptide Y neurons in the hypothalamic dorsomedial nucleus of the rat. Endocrinology 147: 1744-1752.

Larm JA, Gundlach AL (2000) Galanin-like peptide (GALP) mRNA expression is restricted to arcuate nucleus of hypothalamus in adult male rat brain. Neuroendocrinology 72:67-71.

Lawrence CB, Baudoin FM, Luckman SM (2002) Centrally administered galanin-like peptide modifies food intake in the rat: a comparison with galanin. J Neuroendocrinol 14:853-860.

Matsumoto H, Noguchi J, Takatsu Y, Horikoshi Y, Kumano S, Ohtaki T, Kitada C, Itoh T, Onda H, Nishimura O, Fujino M (2001) Stimulation effect of galanin-like peptide (GALP) on luteinizing hormone-releasing hormone-mediated luteinizing hormone (LH) secretion in male rats. Endocrinology 142:3693-3696.

Matsumoto Y, Watanabe T, Adachi Y, Itoh T, Ohtaki T, Onda H, Kurokawa T, Nishimura O, Fujino M (2002) Galanin-like peptide stimulates food intake in the rat. Neurosci Lett 322:67-69.

Ohtaki T, Kumano S, Ishibashi Y, Ogi K, Matsui H, Harada M, Kitada C, Kurokawa T, Onda H, Fujino M (1999) Isolation and cDNA cloning of a novel galanin-like peptide (GALP) from porcine hypothalamus. J Biol Chem 274:37041-37045.

Onaka T, Kuramochi M, Saito J, Ueta Y, Yada T (2005) Galanin-like peptide stimulates vasopressin, oxytocin and adrenocorticotropic hormone release in rats. Neuroreport 16, 243-247.

Parker EM, Izzarelli DG, Nowak HP, Mahle CD, Iben LG, Wang J, Goldstein ME, (1995) Cloning and characterization of the rat GALR1 galanin receptor from Rin14B insulinoma cells. Brain Res Mol Brain Res 34:179-189.

Rodgers RJ, Ishii Y, Halford JC, Blundell JE (2002) Orexins and appetite regulation. Neuropeptides 36:303-325.

Seth A, Stanley S, Dhillo W, Murphy K, Ghatei M, Bloom S (2003) Effects of galanin-like peptide on food intake and the hypothalamo-pituitary-thyroid axis. Neuroendocrinology 77:125-131.

Seth A, Stanley S, Jethwa P, Gardiner J, Ghatei M, Bloom S (2004) Galanin-like peptide stimulates the release of gonadotropin-releasing hormone in vitro and may mediate the effects of leptin on the hypothalamo-pituitary-gonadal axis. Endocrinology 145:743-750.

Takatsu Y, Matsumoto H, Ohtaki T, Kumano S, Kitada C, Onda H, Nishimura O, Fujino M (2001) Distribution of galanin-like peptide in the rat brain. Endocrinology 142:1626-1634.

Takenoya F, Aihara K, Funahashi H, Matsumoto H, Ohtaki T, Tsurugano S, Yamada S, Katoh S, Kageyama H, Takeuchi M, Shioda S (2003) Galanin-like peptide is target for regulation by orexin in the rat hypothalamus. Neurosci Lett 340:209-212.

Takenoya F, Funahashi H, Matsumoto H, Ohtaki T, Katoh S, Kageyama H, Suzuki R, Takeuchi M, Shioda S (2002) Galanin-like peptide is co-localized with alpha-melanocyte stimulating hormone but not with neuropeptide Y in the rat brain. Neurosci Lett 331:119-122.

Takenoya F, Guan JL, Kato M, Sakuma Y, Kintaka Y, Kitamura Y, Kitamura S, Okuda H, Takeuchi M, Kageyama H, Shioda S (2006) Neural interaction between galanin-like peptide (GALP)- and luteinizing hormone-releasing hormone (LHRH)-containing neurons. Peptides 27:2885-2893.

Takenoya F, Hirayama M, Kageyama H, Funahashi H, Kita T, Matsumoto H, Ohtaki T, Katoh S., Takeuchi M, Shioda S (2005) Neuronal interactions between galanin-like-peptide- and orexin- or melanin-concentrating hormone-containing neurons. Regul Pept 126:79-83.

Tatemoto K, Rokaeus A, Jornvall H, McDonald TJ, Mutt V (1983) Galanin - a novel biologically active peptide from porcine intestine. FEBS Lett 164:124-128.

Wang S, Hashemi T, He C, Strader C, Bayne M (1997) Molecular cloning and pharmacological characterization of a new galanin receptor subtype. Mol Pharmacol 52:337-343.

Wittmann G, Sarkar S, Hrabovszky E, Liposits Z, Lechan RM, Fekete C (2004) Galanin- but not galanin-like peptide-containing axon terminals innervate hypophysiotropic TRH-synthesizing neurons in the hypothalamic paraventricular nucleus. Brain Res 1002:43-50.

Functional Analysis of GALP in Feeding Regulation

Haruaki Kageyama[1], Koji Toshinai[2], Yukari Date[3], Masamitsu Nakazato[2] Fumiko Takenoya[1,4], and Seiji Shioda[1]

[1]Department of Anatomy I, Showa University School of Medicine, 1-5-8 Hatanodai, Shinagawa-ku, Tokyo 142-8555, Japan
<e-mail>haruaki@med.showa-u.ac.jp, shioda@med.showa-u.ac.jp
[2]Division of Endocrinology and Metabolism, Department of Internal Medicine, Faculty of Medicine, University of Miyazaki, 5200 Kihara, Kiyotake, Miyazaki 889-1692, Japan
<e-mail>toshinai@med.miyazaki-u.ac.jp, nakazato@med.miyazaki-u.ac.jp
[3]Frontier Science Research Center, University of Miyazaki, 5200 Kihara, Kiyotake, Miyazaki 889-1692, Japan
<e-mail>dateyuka@med.miyazaki-u.ac.jp
[4]Department of Physical Education, Hoshi University School of Pharmacy and Pharmaceutical Science, 2-4-41 Ebara, Shinagawa-ku, Tokyo 142-8501, Japan
<e-mail>kuki@hoshi.ac.jp

Summary. Galanin-like peptide (GALP) is a 60 amino acid neuropeptide isolated from porcine hypothalamus. This peptide, which is produced in neurons within the hypothalamic arcuate nucleus (ARC) of the central nervous system, plays an important role in the regulation of feeding. GALP-containing neurons make an appetite-regulating neuronal circuit with other peptide-containing neurons involved in this process. While centrally administered GALP is known to stimulate acute feeding behavior in the rat, the relevant target neurons have not yet been identified. This review will attempt to summarize the significant body of recent research investigating the ways in which GALP exerts its effects.

Key words. galanin-like peptide (GALP), food intake, lateral hypothalamus, dorsomedial hypothalamus, c-Fos

1 Introduction

Galanin-like peptide (GALP) is a 60 amino acid neuropeptide isolated from porcine hypothalamus. GALP residues 9-21 are identical to the biologically active N-terminal residues 1-13 of galanin (Ohtaki et al. 1999). In vitro, GALP binds with high affinity to galanin receptors 2 (GalR2) and 3 (GalR3) (Kageyama et al. 2005; Lang et al. 2005; Ohtaki et al. 1999).

GALP is produced in the hypothalamus and the posterior pituitary (Fujiwara et al. 2002; Guan et al. 2005; Juréus et al. 2000; Kerr et al. 2000; Larm and Gundlach 2000; Shen et al. 2001; Takatsu et al. 2001). GALP-containing cell bodies are distributed within the arcuate nucleus (ARC), projecting onto orexin- or melanin-concentrating hormone (MCH)-containing neurons in the lateral hypothalamus (LH). GALP-containing fibers are also found in several regions, including the LH, the paraventricular nucleus (PVN), the bed nucleus of the stria terminalis (BST) and the medial preoptic area (MPA) (Takatsu et al. 2001; Takenoya et al. 2006; Takenoya et al. 2005). Within the ARC, the GALP-containing neurons lie in close apposition with Neuropeptide Y (NPY)- and orexin-containing axon terminals (Takenoya et al. 2003; Takenoya et al. 2002). About 10% of these GALP-containing neurons express orexin-1 receptor (Takenoya et al. 2003), and more than 80% express leptin receptor (Takatsu et al. 2001), suggesting that leptin and orexin are involved in the regulation of GALP activity in the hypothalamic neurons.

Many studies have reported that GALP plays a role in the regulation of appetite (Krasnow et al. 2003; Lawrence et al. 2002; Matsumoto et al. 2002), and we have also demonstrated previously that GALP causes an

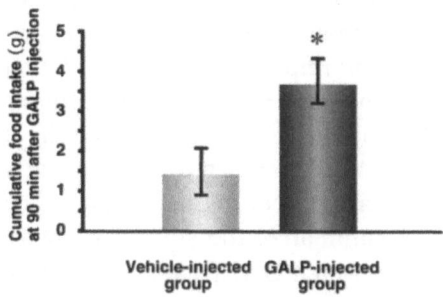

Fig. 1. Cumulative food intake at 90 min after intracerebroventricular infusion of 3 nmol GALP. *$p<0.05$ versus vehicle group.

acute increase in food intake in the rat after being centrally administered (Kageyama et al. 2006) (Fig. 1). However, GALP has also been shown to lead to a reduction in food intake and body weight over a 24-h period (Krasnow et al. 2003; Lawrence et al. 2002; Lawrence et al. 2003). Here, on the basis of our own and other studies, we will review the way in which GALP exerts its effects.

2 c-Fos expression induced by GALP treatment

Many studies investigating the areas of the brain involved in the action of GALP are based on assessment of c-Fos expression, a marker of neuronal activation, after central injection of GALP. The intracerebroventricular (i.c.v.) infusion of GALP activates neurons in several brain regions of the rat. However, different dosages of GALP, the site of injection and nutrient status result in different patterns of c-Fos expression. When GALP is injected into the lateral ventricle, c-Fos-like immunoreactivity is seen in neurons in the MPA, PVN, LH, ARC, supraoptic nucleus (SON) and dorsomedial nucleus of the hypothalamus (DMH), as well as in the brainstem nucleus tractus solitarius (NTS) (Lawrence et al. 2003). Furthermore, GALP activates astrocytes but not microglia in the hypothalamus (Lawrence et al. 2003). In contrast, injection of GALP into the third ventricle induces c-Fos expression in the caudal preoptic area, ARC, median eminence and horizontal limb of the diagonal band of Broca (Fraley et al. 2003). However, the main target area of GALP in the brain and the type of cells affected in the case of increased food intake remain to be determined.

After i.c.v. injection of GALP, nutrient condition, i.e. feeding or fasting, also results in different patterns of c-Fos expression. GALP induces c-Fos expression independent of food intake in the DMH, LH, ARC and periventricular hypothalamus. However, c-Fos expression is not induced in the NTS and SON in rats which are not allowed access to food after injection of GALP (Lawrence et al. 2003). This suggests that c-Fos expression in these areas is a secondary response to food intake, particularly as the SON is a region known to be involved in the maintenance of water balance. It therefore appears that the acute orexigenic effects of centrally administered GALP are mediated in part via the DMH, LH and ARC.

In spite of the above, double immunostaining studies have yet to reveal what types of neurons or glial cells are activated by GALP, and further studies are required to elucidate the physiological significance of GALP in this regard.

3 Neuronal regulation of feeding by GALP

Although centrally administered GALP is known to stimulate feeding behavior (Fig. 1), the target neurons underlying this orexigenic action have not yet been fully established. Recently, both we and others have demonstrated two pathways by which GALP stimulates feeding. One of these is a pathway via orexin neurons in the LH (Kageyama et al. 2006), while the other is via neuropeptide Y neurons in the DMH (Kuramochi et al. 2006) (Fig. 2).

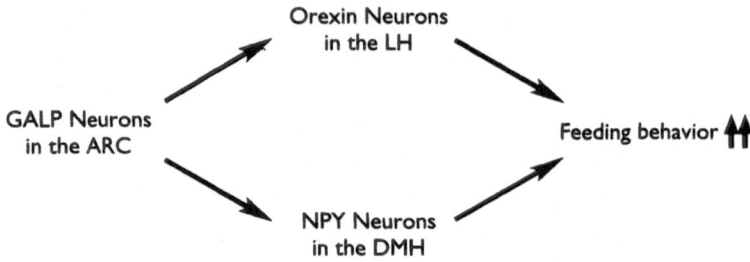

Fig. 2. Regulatory pathway by which GALP stimulates feeding. LH: lateral hypothalamus; DMH: dorsomedial hypothalamus; GALP: galanin-leike peptide; NPY: neuropeptide Y

Previously, we have used a double-immunofluorescence method to show that GALP-immunoreactive fibers lie in direct contact with both orexin- and MCH-immunoreactive neuronal cell bodies in the rat LH (Takenoya et al. 2005). At the ultrastructural level, GALP-immunoreactive axon terminals were found to make synapses on orexin-immunoreactive neuronal cell bodies and dendritic processes in the LH. c-Fos immunoreactivity was expressed in orexin-immunoreactive neurons but not in MCH-immunoreactive neurons in the LH at 90 min after i.c.v. infusion of GALP. Furthermore, the feeding behavior of rats was studied following i.c.v. GALP injection with or without anti-orexin A and B immunoglobulin (IgG) pretreatment. The anti-orexin IgGs markedly inhibited GALP-induced hyperphagia (Kageyama et al. 2006). These results suggest that orexin-containing neurons in the LH are targeted by GALP, and that GALP-induced hyperphagia is mediated mainly via orexin neurons in the rat hypothalamus.

Following GALP injection into the several nuclei in the hypothalamus, food intake has also been measured. GALP administered into the DMN, but not the ARC, LH or PVN, stimulated food intake, whereas central administration of GALP into DMN-lesioned rats produced attenuated feeding compared with that found in vehicle-injected control rats. I.c.v.

infusion of GALP induced c-Fos expression in NPY-containing neurons in the DMN, as well as increasing the cytosolic calcium concentration in these cells. However, no such effects were observed in the ARC. Furthermore, pre-injection of both anti-NPY IgG and NPY antagonists have been reported to attenuate GALP-induced hyperphagia (Kuramochi et al. 2006). These results suggest that i.c.v. infusion of GALP stimulates food intake via activation of NPY-containing neurons in the DMN. However, because GALP-immunoreactive fibers have not been reported in the DMN, it remains unclear whether GALP transmits to NPY-containing neurons via synapses between GALP and NPY neurons in vivo. Recently, it was reported that intra-MPA or intra-PVN injection of GALP potently increases cumulative food intake in rats over the first 1 h post-treatment (Patterson et al. 2006). However, the type of neurons activated in these nuclei remains unclear. Further neuronal anatomical studies are required in order to understand the precise mechanisms by which GALP regulates feeding behavior.

Acknowledgement

The authors thank Dr. Tetsuya Ohtaki in Takeda Chemical Industrial Company and Ms. Sachi Kato for their help to accomplish this study. This study was supported in part by a grant from the Ministry of Education, Culture, Sports, Science and Technology of Japan.

References

Fraley GS, Shimada I, Baumgartner JW, Clifton DK, and Steiner RA (2003). Differential patterns of fos induction in the hypothalamus of the rat following central injections of galanin-like Peptide and galanin. Endocrinology 144:1143-1146.

Fujiwara K, Adachi S, Usui K, Maruyama M, Matsumoto H, Ohtaki T, Kitada C, Onda H, Fujino M, and Inoue K (2002). Immunocytochemical localization of a galanin-like peptide (GALP) in pituicytes of the rat posterior pituitary gland. Neurosci Lett 317:65-68.

Guan JL, Kageyama H, Wang QP, Takenoya F, Kita T, Matsumoto H, Ohtaki T, and Shioda S (2005). Electron microscopy examination of galanin-like peptide (GALP)-containing neurons in the rat hypothalamus. Regul Pept 126:73-78.

Juréus A, Cunningham MJ, McClain ME, Clifton DK, and Steiner RA (2000). Galanin-like peptide (GALP) is a target for regulation by leptin in the hypothalamus of the rat. Endocrinology 141:2703-2706.

Kageyama H, Kita T, Toshinai K, Guan JL, Date Y, Takenoya F, Kato S,

Matsumoto H, Ohtaki T, Nakazato M, and Shioda S (2006). Galanin-like peptide promotes feeding behaviour via activation of orexinergic neurones in the rat lateral hypothalamus. J Neuroendocrinol 18:33-41.

Kageyama H, Takenoya F, Kita T, Hori T, Guan JL, and Shioda S (2005). Galanin-like peptide in the brain: effects on feeding, energy metabolism and reproduction. Regul Pept 126:21-26.

Kerr NC, Holmes FE, and Wynick D (2000). Galanin-like peptide (GALP) is expressed in rat hypothalamus and pituitary, but not in DRG. Neuroreport 11:3909-3913.

Krasnow SM, Fraley GS, Schuh SM, Baumgartner JW, Clifton DK, and Steiner RA (2003). A role for galanin-like peptide in the integration of feeding, body weight regulation, and reproduction in the mouse. Endocrinology 144:813-822.

Kuramochi M, Onaka T, Kohno D, Kato S, and Yada T (2006). Galanin-like peptide stimulates food intake via activation of neuropeptide Y neurons in the hypothalamic dorsomedial nucleus of the rats. Endocrinology 147:1744-1752.

Lang R, Berger A, Santic R, Geisberger R, Hermann A, Herzog H, and Kofler B (2005). Pharmacological and functional characterization of galanin-like peptide fragments as potent galanin receptor agonists. Neuropeptides 39:179-184. Epub 2005 Feb 2001.

Larm JA, and Gundlach AL (2000). Galanin-like peptide (GALP) mRNA expression is restricted to arcuate nucleus of hypothalamus in adult male rat brain. Neuroendocrinology 72:67-71.

Lawrence CB, Baudoin FM, and Luckman SM (2002). Centrally administered galanin-like peptide modifies food intake in the rat: a comparison with galanin. J Neuroendocrinol 14:853-860.

Lawrence CB, Williams T, and Luckman SM (2003). Intracerebroventricular galanin-like peptide induces different brain activation compared with galanin. Endocrinology 144:3977-3984.

Matsumoto Y, Watanabe T, Adachi Y, Itoh T, Ohtaki T, Onda H, Kurokawa T, Nishimura O, and Fujino M (2002). Galanin-like peptide stimulates food intake in the rat. Neurosci Lett 322:67-69.

Ohtaki T, Kumano S, Ishibashi Y, Ogi K, Matsui H, Harada M, Kitada C, Kurokawa T, Onda H, and Fujino M (1999). Isolation and cDNA cloning of a novel galanin-like peptide (GALP) from porcine hypothalamus. J Biol Chem 274:37041-37045.

Patterson M, Murphy KG, Thompson EL, Smith KL, Meeran K, Ghatei MA, and Bloom SR (2006). Microinjection of galanin-like peptide into the medial preoptic area stimulates food intake in adult male rats. J Neuroendocrinol 18:742-747.

Shen J, Larm JA, and Gundlach AL (2001). Galanin-like peptide mRNA in neural lobe of rat pituitary. Increased expression after osmotic stimulation suggests a role for galanin-like peptide in neuron-glial interactions and/or neurosecretion. Neuroendocrinology 73:2-11.

Takatsu Y, Matsumoto H, Ohtaki T, Kumano S, Kitada C, Onda H, Nishimura O, and Fujino M (2001). Distribution of galanin-like peptide in the rat brain. Endocrinology 142:1626-1634.

Takenoya F, Aihara K, Funahashi H, Matsumoto H, Ohtaki T, Tsurugano S,

Yamada. S, Katoh S, Kageyama H, Takeuchi M, and Shioda S (2003). Galanin-like peptide is target for regulation by orexin in the rat hypothalamus. Neurosci Lett 340:209-212.

Takenoya F, Funahashi H, Matsumoto H, Ohtaki T, Katoh S, Kageyama H, Suzuki R, Takeuchi M, and Shioda S (2002). Galanin-like peptide is co-localized with alpha-melanocyte stimulating hormone but not with neuropeptide Y in the rat brain. Neurosci Lett 331:119-122.

Takenoya F, Guan JL, Kato M, Sakuma Y, Kintaka Y, Kitamura Y, Kitamura S, Okuda H, Takeuchi M, Kageyama H, and Shioda S (2006). Neural interaction between galanin-like peptide (GALP)- and luteinizing hormone-releasing hormone (LHRH)-containing neurons. Peptides 27:2885-2893. Epub 2006 Jun 2821.

Takenoya F, Hirayama M, Kageyama H, Funahashi H, Kita T, Matsumoto H, Ohtaki T, Katoh S, Takeuchi M, and Shioda S (2005). Neuronal interactions between galanin-like-peptide- and orexin- or melanin-concentrating hormone-containing neurons. Regul Pept 126:79-83.

Regulation of Energy Homeostasis by GALP

Seiji Shioda[1], Haruaki Kageyama[1], Fumiko Takenoya[1, 2], Yukari Date[3], Masamitsu Nakazato[3], Toshimasa Osaka[4], Yasuhiko Minokoshi[5]

[1]Department of Anatomy I, Showa University School of Medicine, 1-5-8 Hatanodai, Shinagawa-ku, Tokyo 142-8555, Japan
<e-mail> Shioda@med.showa-u.ac.jp, haruaki@med.showa-u.ac.jp
[2]Department of Physical Education, Hoshi University School of Pharmacy and Pharmaceutical Science, 2-4-41 Ebara, Shinagawa, Tokyo 142-8501, Japan
<e-mail> kuki@hoshi.ac.jp
[3]Frontier Science Research Center, University of Miyazaki, 5200 Kihara, Kiyotake, Miyazaki 889-1692, Japan
<e-mail> dateyuka@med.miyazaki-u.ac.jp
[4]Division of Endocrinology and Metabolism, Department of Internal Medicine, Faculty of Medicine, University of Miyazaki, 5200 Kihara, Kiyotake, Miyazaki 889-1692, Japan
<e-mail> nakazato@med.miyazaki-u.ac.jp
[5]National Institute of Health and Nutrition, 1-23-1 Toyama, Shinjuku-ku, Tokyo 162-8636, Japan
<e-mail> osaka@nih.go.jp
[5]National Institutes for Physiological Sciences, 38 Nishigonaka Myodaiji, Okazaki, Aichi 444-8585, Japan
<e-mail> minokosh@nips.ac.jp

Summary. The hypothalamus plays very important roles in regulation of feeding behavior, energy metabolism and reproduction. Galanin-like peptide (GALP) was discovered in porcine hypothalamus and has 60 amino acid peptide and a non-amidated C-terminus. GALP is shown to be produced in neurons in the hypothalamic arcuate nucleus. GALP-producing neurons make neuronal networks with several feeding-related peptide-containing neurons in the hypothalamus. Since GALP controls food intake and energy balance, it is assumed that GALP is an important neuropeptide to regulate body weight. Furthermore, GALP is known to regulate plasma LH levels through activation of gonadotrophin-releasing hormone (GnRH)-producing neurons in the hypothalamus, suggesting that GALP plays some important roles in the reproductive system. This review will attempt to summarize the research on these topics especially on energy metabolism and reproduction.

Key words. Galanin-like peptide, Feeding, Energy metabolism,

Neuronal network

1. Introduction

Galanin, isolated from porcine intestine, is a 29 amino acid peptide and its receptors have 3 subtypes, GALR1, GALR2 and GALR3 (Kageyama et al, 2005). Since galanin-like peptide (GALP) has been discovered as an endogenous ligand for galanin receptor 2, several studies have shown physiological functions of GALP. GALP is isolated from the porcine hypothalamus (Ohtaki et al, 1999) on the basis of its ability to bind and activate galanin receptors in vitro. GALP is a novel 60 amino acid peptide and its amino acid residues (9-21) are identical with the biologically active N-terminal (1-13) portion of galanin (Ohtaki et al, 1999). GALP cDNA is cloned from porcine, rat, mouse, monkey and human (Kageyama et al, 2005).

Here, we will review a number of studies that advance in our understanding the functional significance of energy homeostasis and reproductive system and suggest possible future research that may help clarify the precise roles of GALP on these systems.

2. Feeding behavior and energy metabolism

Intracerebroventricular (icv) injection of GALP into the lateral ventricle increases food intake in the first 1-2 h (Matsumoto et al, 2002; Lawrence et al, 2003; Seth et al, 2003). The potency of GALP is ten times higher than that of galanin for the stimulation of food intake (Matsumoto et al, 2002; Lawrence et al, 2002). There is species difference on GALP-induced feeding behavior: mouse vs. rat. In mouse, icv injection of GALP into the ventricle decreases in food intake at the first 1 h (Krasnow et al, 2003). Body weight gain and food intake are reduced at 24-h post injection. Icv injection of GALP into the lateral ventricle of *ob/ob* mouse decreases in both food intake and body weight gain at early time (Hansen et al, 2003). In a pair-fed study, chronic GALP administration induces a greater decrease in body weight (Hansen et al, 2003). We have recently shown that GALP promotes feeding behavior via activation of orexin-containing neurons in the lateral hypothalamus (Kageyama et al, 2006). These results indicate that GALP has an acute orexigenic action on feeding and anorectic action at a later time.

It is shown that chronic GALP injection lasts to decrease in body weight and increases in core body temperature, despite recovery of food intake

(Hansen et al, 2003). Core body temperature is increased in a dose-dependent manner and then lasts until 6-8 h post injection (Lawrence et al, 2002; Krasnow et al, 2003). We have observed when rats are given an icv injection of 0, 0.5, 1 or 2.5 nmol GALP they show upregulation of O_2 consumption, heart rate and core body temperature in a dose-dependent manner. However, iv injection of GALP at high dose, 2.5 nmol, there are no significant changes in O_2 consumption, heart rate, or core body temperature (Kageyama et al in preparation). These results suggest that GALP-induced heat production is a process mediated in the central nervous system (CNS).

The precise mechanisms of heat production by GALP in the CNS is still not yet known. We have preliminary data that GALP activates astrocytes which in turn increases cyclooxygenase activity and then stimulates prostaglandin production. As prostaglandin acts on the hypothalamic preoptic area, and it elevates temperature, it appears that GALP acts on astrocytes to produce prostaglandin following body temperature. We are now trying to make another experiment to determine whether GALP affects glucose uptake in skeletal muscles in rat. The brown adipose tissue (BAT) is closely related to heat production in rodents and uncoupling protein (UCP)-1 is expressed in BAT. It is also reported that GALP treatment increases both UCP-1 mRNA expression and protein in BAT of *ob/ob* mouse (Hansen et al, 2003). The UCP-1 in the BAT is regulated by the sympathetic nervous system (Onai et al, 1995). The number of proopiomelanocortin (POMC) mRNA-expressing cells in arcuate nucleus of ob/ob mouse is reduced after chronic GALP injection (Hansen et al, 2003). These findings suggest that activation of sympathetic nervous system and thermogenesis by leptin may be partially mediated by GALP through a POMC-independent mechanism. Recently, polysaccaride is reported to induce GALP mRNA in rat, suggesting that GALP is related to inflamed thermogenesis (Saito et al, 2003, 2005). Our and other studies may suggest that GALP plays very important roles in energy metabolism as well as body weight regulation through the sympathetic nervous system (Fig. 1).

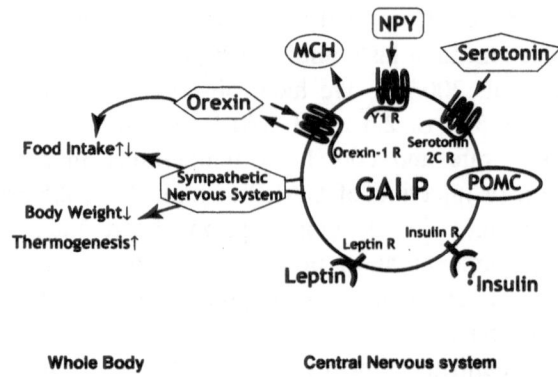

Fig. 1.   Mechanism of GALP on feeding regulation and thermoregulation.

3. The regulation of reproductive function

GALP-immunoreactive fibers are in close apposition with GnRH-immunoreactive fibers in the medial preoptic area (MPA) and bed nucleus of the stria terminalis (BST) and 6 % of GnRH-immunoreactive neurons in the MPA show close contact with GALP-immunoreactive fibers (Takatsu et al, 2001). In rats, icv infusion of GALP increases the plasma luteinizing hormone (LH) level between 10 and 60 min postinjection but does not change the levels of other hormones (Matsumoto et al, 2001). Central injection of GALP is shown to increase the LH secretion in the macaque and it is blocked by administration of the GnRH receptor antagonist (Cunningham et al, 2004). Centrally administration of GALP is reported to induce c-Fos activation in GnRH neurons in the in the MPA (Matsumoto et al, 2001).

On the other hand, in mice, icv injection of GALP increases in plasma LH level and plasma testosterone levels but not plasma follicle-stimulating hormone (FSH) level, similar to rats (Krasnow et al, 2003). In addition, treatment with GALP stimulates the release of GnRH from hypothalamus explants but also the immortalized GnRH cell line, GT1-7 cells. However, RT-PCR revealed that none of galanin receptors is expressed in GT1-7 cells (Seth et al, 2004). Therefore, there are possible two pathways on the release of GnRH via GALP. One is a direct activation of GnRH neurons via a novel receptor. Another is an indirect pathway through galanin receptor-expressed inter neurons.

To detect the presence of GALP- and GnRH-containing neurons in the brain, we studied neuronal interactions between GALP- and GnRH-containing neurons in the rat hypothalamus at the ultrastructural level. We used transgenic rats that expressed green fluorescent dye (GFP) predominantly in GnRH neurons (Takenoya et al, 2006). In these animals, GFP- and GnRH-like immunoreactivity was seen to overlap each other in the median eminence, organum vasculosum of the lamina terminalis, and MPA. This may indicate that this transgenic rat model is suitable for examining neural interactions between GALP and GnRH by morphological methods. In the hypothalamic

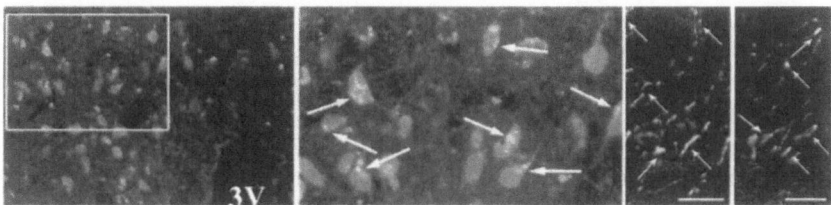

Fig. 2 Immunofluorescence photomicrographs of sections with direct dual-labeling combining antisera to GALP and LHRH. A; Immunoreactivity for GALP (red) and LHRH (green) is localized in the MPA respectively. B; The merged images show some GALP-positive terminals in direct contact (arrows) with LHRH neuronal cell bodies and processes. C; Immunoreactivity for GALP (red) and LHRH (green) is localized in the BST. Some GALP- positive terminals in direct contact with LHRH neuronal processes (arrows). Scale bars =20 μm. 3V: third ventricle

preoptic area, a double-immunostaining method revealed that GALP-positive fibers were in direct contact with GnRH-containing neurons (Takenoya et al, 2006). Double-immunoelectron micrographs also revealed that GALP-positive axon terminals made synaptic contact with GnRH-containing cell bodies and dendritic processes in the preoptic area (Takenoya et al, 2006). These results strongly suggest that GALP neurons make direct input to GnRH neurons and GALP plays a crucial role in the regulation of LH secretion via GnRH.

Acknowledgments

The authors thanks Dr. Tetsuya Ohtaki in Takeda Chemical Industrial Company and Ms. Sachi Kato for their help to accomplish this study. This study was supported in part by the High-Technology Research Center Project and Showa University Grant-in Aid for Innovative Collaborative Research Projects and a Special Research Grant-in Aid for Development of

Characteristic Education from the Japanese Ministry of Education, Sports, Science and Technology.

References

Cunningham MJ, Shahab M, Grove KL, Scarlett JM, Plant TM, Cameron JL, Smith MS, Clifton DK, Steiner RA (2004) Galanin-like peptide as a possible link between metabolism and reproduction in the macaque. J Clin Endocrinol Metab 89:1760-1766

Hansen KR, Krasnow SM, Nolan MA, Fraley GS, Baumgartner JW, Clifton DK, Steiner RA (2003) Activation of the sympathetic nervous system by galanin-like peptide--a possible link between leptin and metabolism. Endocrinology 144:4709-4717

Kageyama H, Takenoya F, Kita T, Hori T, Guan JL, Shioda S (2005) Galanin-like peptide in the brain: effects on feeding, energy metabolism and reproduction. Regul Pept 126:21-26

Kageyama H, Kita T, Toshinai K, Guan JL, Date Y, Takenoya F, Nakazato M, Shioda S (2006) Galanin-like peptide promotes feeding behavior via activation of orexinergic neurons in the rat lateral hypothalamus. J Neuroendocrinol 18:33-41

Krasnow SM, Fraley GS, Schuh SM, Baumgartner JW, Clifton DK, Steiner RA (2003) A role for galanin-like peptide in the integration of feeding, body weight regulation, and reproduction in the mouse. Endocrinology 144: 813-822

Lawrence CB, Baudoin FM, Luckman SM (2002) Centrally administered galanin-like peptide modifies food intake in the rat: a comparison with galanin. J Neuroendocrinol 14: 853-860

Lawrence CB, Williams T, Luckman SM (2003) Intracerebroventricular galanin-like peptide induces different brain activation compared with galanin. Endocrinology 144:3977-3984

Matsumoto H, Noguchi J, Takatsu Y, Horikoshi Y, Kumano S, Ohtaki T, Kitada C, Itoh T, Onda H, Nishimura O, Fujino M (2001) Stimulation effect of galanin-like peptide (GALP) on luteinizing hormone-releasing hormone-mediated luteinizing hormone (LH) secretion in male rats. Endocrinology 142:3693-3696

Matsumoto Y, Watanabe T, Adachi Y, Itoh T, Ohtaki T, Onda H, et al. Galanin-like peptide stimulates food intake in the rat. Neurosci Lett 2002; 322: 67-69.

Ohtaki T, Kumano S, Ishibashi Y, Ogi K, Matsui H, Harada M, Kitada C, Kurokawa T, Onda H, Fujino M (1999) Isolation and cDNA cloning of a novel galanin-like peptide (GALP) from porcine hypothalamus. J Biol Chem 274: 37041-37045

Onai T, Kilroy G, York DA, Bray GA (1995) Regulation of beta 3-adrenergic receptor mRNA by sympathetic nerves and glucocorticoids in BAT of Zucker obese rats. Am J Physiol 269:R519-R526

Saito J, Ozaki Y, Ohnishi H, Nakamura T, Ueta Y (2003) Induction of galanin-like peptide gene expression in the rat posterior pituitary gland during endotoxin shock and adjuvant arthritis. Brain Res Mol Brain Res 113:124-132

Saito J, Ozaki Y, Kawasaki M, Ohnishi H, Okimoto N, Nakamura T, Ueta Y (2005) Induction of galanin-like peptide gene expression in the arcuate nucleus of the rat after acute but not chronic inflammatory stress. Brain Res Mol Brain Res 133:233-241

Seth A, Stanley S, Dhillo W, Murphy K, Ghatei M, Bloom S (2003) Effects of galanin-like peptide on food intake and the hypothalamo-pituitary-thyroid axis. Neuroendocrinology 77:125-131

Seth A, Stanley S, Jethwa P, Gardiner J, Ghatei M, Bloom S (2004) Galanin-like peptide stimulates the release of gonadotropin-releasing hormone in vitro and may mediate the effects of leptin on the hypothalamo-pituitary-gonadal axis. Endocrinology   145:743-750.

Takatsu Y, Matsumoto H, Ohtaki T, Kumano S, Kitada C, Onda H, Nishimura O, Fujino M (2001) Distribution of galanin-like peptide in the rat brain. Endocrinology 142:1626-1634

Takenoya F, Guan JL, Kato M, Sakuma Y, Kintaka Y, Kitamura Y, Kitamura S, Okuda H, Takeuchi M, Kageyama H, Shioda S (2006) Neural interaction between galanin-like peptide (GALP)- and luteinizing hormone-releasing hormone (LHRH)-containing neurons. Peptides 27:2885-2893

Part II
Lipid Metabolism and Atherosclerosis

Cholesterol Trafficking and Esterification With Relation to Atherosclerosis and Neurodegenerative Diseases

Ta-Yuan Chang and Catherine C.Y. Chang

Department of Biochemistry, Dartmouth Medical School, Hanover, NH, 03755 USA
e-mail: Ta.Yuan.Chang@dartmouth.edu

Summary: Cholesterol is an important lipid molecule that is needed for the growth and viability of mammalian cells. The metabolites of cholesterol include bile acids, oxysterols, and steroid hormones; these substances also have important physiological functions. The sources of cholesterol are from the diet, and from endogenous biosynthesis. Genetically inherited and diet-induced hypercholesterolemia are major risk factors for causing atherosclerotic cardiovascular disease. This Chapter reviews our current knowledge on cellular cholesterol sources and trafficking routes in a mammalian cell. It also reviews the biochemical properties, and the cell biological and pathophysiological roles of the enzyme acyl coenzyme A:cholesterol acyltransferase (ACAT).

Key words: Cholesterol trafficking, cholesterol metabolism, atherosclerosis, Niemann-Pick Type C disease, neurodegenerative diseases

I. Cellular cholesterol sources and trafficking routes in a mammalian cell

Three processes govern the net cholesterol content in a single mammalian cell:

1. Cellular cholesterol input

Cells receive cholesterol mainly from low-density lipoprotein (LDL), the major cholesterol carrier in the blood, via receptor-mediated endocytosis

(Brown and Goldstein 1986). Cells also synthesize cholesterol de novo from the simple precursor acetyl coenzyme A. Biosynthesis of cholesterol involves the actions of numerous enzymes. The integral membrane protein complex SCAP/SREBP, located at the ER, controls the expression of many genes involved in cholesterol biosynthesis (reviewed in (Goldstein et al. 2006)). SREBPs are transcription factors. The precursor forms of SREBPs are integral membrane proteins residing in the ER. SCAP, a multi-span membrane protein with a sterol-sensing domain (SSD), binds to the SREBP in the ER with high affinity. The decrease in cellular cholesterol level allows the SCAP/SREBP to leave ER and be transported to the Golgi via the secretory pathway. Upon arriving at the Golgi, the SREBPs are cleaved sequentially by two proteases called S1P and S2P. The cleavages produce the soluble, transcriptionally active N-terminal domains of SREBPs, which then enter the nucleus and act as mature transcription factors in activating many genes involved in lipid biosyntheses, including the key enzyme HMG-CoA reductase (HMGR), another multi-span membrane protein with SSD, located in the ER.

2. Cellular cholesterol output

The output is governed by a process called reverse cholesterol transport, a metabolic pathway whereby excess cholesterol in peripheral tissues is transported to the liver for elimination from the body or for reutilization. Cells release cholesterol at the plasma membranes (PM) to high-density lipoproteins (HDLs). Cells also release cholesterol and phospholipids to helical apoproteins, including apo-A1, apoE, etc., through the lipid efflux protein ABCA1 located at the PM (discussed in (Chang et al. 2006)).

3. The cholesterol/cholesteryl ester cycle

The enzyme acyl coenzyme A:cholesterol acyltransferase (ACAT) converts cholesterol to cholesteryl esters, which are stored as cytoplasmic lipid droplets. The enzyme cholesteryl ester hydrolase(s) convert cholesteryl esters back to cholesterol. The CHOL/cholesteryl ester cycle continues dynamically within the cell. The half-time of CE hydrolysis is several hours. CHOL released from the CE may be an efficient substrate for apoA1-mediated CHOL efflux (discussed in (Chang et al. 2006)).

Fate of LDL-derived CHOL

To monitor the fate of lipid cargo cholesteryl esters (CE) present in LDL, one typically uses radiolabeled cholesteryl ester incorporated into LDL, and monitors the radioactive cholesterol released from CE. The results of pulse-chase experiments show that CE in LDL undergoes hydrolysis in approximately 30 min in a unique, early endosomal-like compartment enriched in the enzyme acid lipase (Sugii et al. 2003). The cholesterol released from the acid lipase compartment then emerges in the late endosomes (Wojtanik and Liscum 2003), (Sugii et al. 2003). Within another 30 min, the LDL-derived cholesterol (LDL-CHOL) egresses from the late endosomes and arrives at the PM (Sugii et al. 2003). PM cholesterol is available to extraction by cyclodextrin, a water-soluble, membrane-impermeable molecule that binds to cholesterol with high affinity. The arrival of radiolabeled, LDL-CHOL at the PM can be monitored by adding cyclodextrin to the growth medium for 5 or 10 min, then counting the radioactivity present in cyclodextrin. LDL-CHOL also arrives at the ER, and becomes available for esterification by the resident ER enzyme ACAT1; the amount of radioactive LDL-CHOL re-esterified can be measured by thin-layer chromatographic analysis (Cruz et al. 2000). The egress of LDL-CHOL from late endosomes to the PM for sterol efflux and to the ER for re-esterification critically depends on a pair of proteins: Niemann-Pick type C1 (NPC1) and NPC2. These proteins are named after a rare disease called the Niemann-Pick type C disease (reviewed in (Patterson et al. 2001), (Liscum and Sturley 2004), (Chang et al. 2005)). Niemann-Pick type C disease (NPC) is a rare, genetic, neurodegenerative disorder. It is responsible for the build-up of cholesterol, gangliosides, and other lipids in the spleen, liver, and brain. The build-up of lipids results in neurodegeneration in the central nervous system. The disease is fatal to children who carry two mutant copies of the NPC gene. Currently, there is no cure for this disease. Two different genes can cause the disease, *NPC1* and *NPC2*. Mice models for NPC1 and NPC2 disease have become available, and have served as valuable research tools (Loftus et al. 1997), (Sleat et al. 2004). In mutant NPC1 and NPC2 cells, the egress of cholesterol, and possibly other lipids from the late endosomes/lysosomes, is severely defective. NPC1 is a multi-span membrane protein that contains the SSD, and is believed to be located in the late endosomes. The NPC2 protein is a soluble, glycosylated protein located in the lumen of the late endosomes/lysosomes, and can be secreted into the medium. Both proteins bind to cholesterol (Ohgami et al. 2004), (Naureckiene et al. 2000). NPC2 proteins are believed to transport cholesterol within the late endosomes/lysosomes (Cheruku et al. 2006). Cholesterol accumulation in NPC1 and NPC2 mutants can be prevented by overexpressing the Rab9 protein or other Rab proteins (Choudhury et al. 2002), (Walter et al. 2003).

Rab9 is one of the key late endosomal-specific small GTPases that play important roles in vesicular trafficking, including membrane vesicle docking, fusion, etc., especially between the late endosomes and the trans-Golgi network (Pfeffer and Aivazian 2004). The fact that the NPC phenotype could be abolished by overexpressing the Rab proteins suggests that within the late endosomal system, there may be two distinct mechanisms for cholesterol transport: one mediated by NPC, the other involving a Rab-dependent vesicular trafficking mechanism which can bypass the function of NPC.

Fate of endoSTEROLs

The late stage of cholesterol synthesis takes place at the ER. To monitor the fate of endogenously synthesized sterols (designated as endoSTEROLS), one typically feeds intact cells with radioactive acetate, the simple precursor for *de novo* cholesterol biosynthesis, for a short period of time (within minutes). The amount of radioactive CHOL and its biosynthetic precursor sterols can be measured by thin-layer chromatographic analysis (Yamauchi et al., results to be published). The results of pulse-chase experiments show that, once biosynthesized, most if not all of the endoCHOL is rapidly transported to the cholesterol-rich, sphingolipid-rich domains (i.e., lipid rafts/caveolae) of the PM within 10 to 20 min, by an energy-dependent, non-vesicular trafficking process. This process involves caveolin1 and other soluble proteins (reviewed in (Matveev et al. 2001)), and is independent of NPC1 (Liscum et al. 1989). Within 2-4 hrs after its initial synthesis, endoCHOL moves to late endosomes (Sugii et al. 2006). After several additional hours, endoCHOL accumulates in the (aberrant) late endosomes/lysosomes of mutant NPC1 cells, but not those of the parental cells (Lange et al. 1998), (Cruz and Chang 2000). In the mutant NPC1 cells, the movements of endoCHOL from the late endosomes/lysosomes back to the PMs, and to the ER for esterification, are both partially defective.

Cell type specificity and redundancy in cholesterol transport

Initially, the endoCHOL trafficking defects described above were demonstrated by comparing the properties of a mutant NPC1 cell line CT43 vs. its parental CHO cell line 25RA. 25RA is a CHO cell line resistant to the cytotoxicity of 25-hydroxycholesterol, and resistant to sterol-mediated downregulation (Chang and Limanek 1980), (Hua et al.,

1997). It contains a gain-of-function mutation in the SREBP cleavage activating protein (SCAP). The CT43 mutant cell line is isolated as one of the cholesterol trafficking mutants from mutagenized 25RA cells (Cadigan et al. 1990). It contains the same gain-of-function mutation in the protein SCAP. In addition, it contains a premature translational termination mutation near the 3'-end of the NPC1 coding sequence, producing a non-functional, truncated NPC1 protein (Cruz et al., 2000). Thus, it was not clear whether the results obtained by using these cells were applicable to other cells types. To address this issue, additional experiments were performed in parallel, in four different cell types: embryonic fibroblasts, hepatocytes, cerebellar glial cells, and peritoneal macrophages; the cells were all isolated from the NPC1 mice and their wild-type littermates. The use of cells isolated from the wild-type littermate mice to serve as the parental cells avoids the possible genetic heterogeneity that may exist between the parental cells and the mutant NPC1 cells. The results show the endoCHOL trafficking defects can be easily demonstrated in mutant macrophages and mutant glial cells, but cannot be easily demonstrated in mutant fibroblasts or mutant hepatocytes (Reid et al. 2003). Thus, the dependency on NPC1 for endoCHOL trafficking is cell-type specific. The cell-type-specific dependency may be explained by the following scenario: EndoCHOL first arrives at the PM. A certain portion of endo-CHOL internalizes from the PM to the interior of the late endosomes/lysosomes by a bulk PM endocytosis process, such that its egress becomes dependent on NPC1. Another portion of endoCHOL undergoes retrograde transport from the PM to the ER, without the participation of NPC1/NPC2. The retrograde PM to ER cholesterol transport may involve a non-clathrin-mediated, lipid raft/caveolae-mediated endocytic pathway (Yamauchi et al., results to be published), (Pipalia et al. 2007). Other results also show that within a single cell, multiple vesicular and nonvesicular cholesterol transport systems exist (Hao et al. 2002). The molecular nature of any of these pathways described above is not well-understood at present. The apparent redundancy in cholesterol transport may be to assure dynamic movement of cholesterol among various cellular membranes.

II. Acyl coenzyme A:cholesterol acyltransferase (ACAT)

The ACAT-like enzyme family

The first *Acat* gene (*Acat1*) was cloned in 1993 (Chang et al. 1993). The ACAT-like enzyme family is comprised of ACAT1, ACAT2, acyl-coenzyme A:diacylglycerol acyltransferase 1 (DGAT1) (Buhman et al. 2000), and many other enzymes of diverse biological functions (Hofmann 2000). In mammals, the expression of ACAT1 is ubiquitous, while the expression of ACAT2 is mainly expressed in the intestinal enterocytes and liver hepatocytes (Parini et al. 2004), (Song et al. 2006). Both enzymes are potential drug targets for therapeutic intervention against atherosclerosis and hyperlipidemia. ACAT1 is an integral membrane protein present in the ER, with a 50 kDa apparent molecular weight on SDS-PAGE. Due to the difficulty of crystallizing membrane proteins, the structural models of ACAT1 and related proteins are not currently available. Unlike many other enzymes/proteins involved in cellular cholesterol metabolism, ACAT1 is not regulated at the transcriptional level by the cholesterol-dependent SCAP/SREBP pathway. Instead, the activity of ACAT1 is allosterically activated by one of its substrates, cholesterol (reviewed in (Chang et al. 2006)). ACAT1 is a homotetrameric enzyme (Yu et al. 1999) and contains nine transmembrane domains, with the active site (His 460) located within TMD7 (Guo et al. 2005). A hydrophilic segment near the N-terminal located on the cytoplasmic side of the ER membrane contains one dimerization domain; the deletion of which converts the enzyme to a fully functional homodimeric form (Yu et al. 2002). The human ACAT2 encodes a single 46-kDa protein on SDS-PAGE. It is also an integral membrane protein (Lin et al. 2003), and shares high homology with ACAT1 near the C-terminus but not near the N-terminus. ACAT2 may also be allosterically regulated by cholesterol (Liu et al. 2005).

The Acat1 KO and Acat2 KO mice

Farese and colleagues created both the *Acat1-/-* and the *Acat2-/-* mice (Farese 1998), which have served as important research tools in lipoprotein and atherosclerosis research.

ACAT inhibitors

Several pharmaceutical companies have produced specific ACAT inhibitors that have been tested in vitro, in intact cells, and in animals. Whether ACAT inhibitors will serve as effective anti-atherosclerosis drugs is under much debate (see (Leon et al. 2005) and (Nissen et al. 2006) for two different opinions).

Relationship between ACAT1, cholesterol storage, cholesterol efflux, LCAT, HDL, and atherosclerosis

In early stages of the disease atherosclerosis, injuries occur to the lining cells of the artery, causing circulating monocytes to adhere to the injury sites. The monocytes migrate underneath the lining cells of the artery, differentiate into macrophages, and act as scavenging cells to remove various cytotoxic substances. In the macrophages, chronic exposure of cholesterol-rich substances continuously stimulates ACAT1, the major isoenzyme present in macrophages (Miyazaki et al. 1998), to esterify cholesterol. In addition, ACAT1 in macrophages efficiently converts endogenously synthesized cholesterol into cholesteryl ester droplets (Klansek et al. 1996). The action of ACAT1 causes macrophages to become foam cells. The foam cell is a hallmark of early atherosclerotic lesions. Foam cells develop into lipid-rich, vulnerable plaques that may be prone to rupture and blood clot within the artery (reviewed in (Libby and Aikawa 2002)). The cholesterol burden in macrophages can be partially relieved by blocking endogenous sterol synthesis, and/or by the reverse cholesterol transport process (RCT) that involves efflux of plasma membrane cholesterol via several cellular transporters (ABCA1, ABCG1, SRB1, etc.) to various forms of HDL. The effluxed cholesterol is then converted to cholesteryl esters by the enzyme lecithin:cholesterol acyltransferase (LCAT); the CEs in HDL are then selectively removed by the hepatic scavenger receptor SRB1 (discussed in (Chang et al. 2006). Macrophage reverse cholesterol transport may play a pivotal role in regression and prevention of atherosclerotic lesions (reviewed in (Cuchel and Rader 2006)).

Cholesterol, ACAT inhibitor, and Alzheimer's disease

Alzheimer's disease (AD) is characterized by massive extracellular accumulation of amyloid plaques, mainly composed of amyloid beta (A-beta) peptide aggregates, and intracellular accumulation of hyperphosphorylated tau protein in the cerebellar area. These events lead to widespread neurodegeneration in the brain. Genetic variation in apolipoprotein (apo) E, a plasma CHOL transport protein, is a risk factor for AD (Raffai and Weisgraber 2003), (Ye et al. 2005). In addition, studies suggest that CHOL in membranes impacts on production of A-beta (reviewed in (Wolozin 2004)). Amyloid precursor protein (APP) can be cleaved via two competing pathways, the alpha and the beta secretase pathways, which are distinguished by the subcellular site of proteolysis and the site of cleavage within APP. Several proteases are capable of producing the alpha-cleavage after which the gamma-secretase complex, which includes presenilin 1 as a catalytic subunit, further cleaves the APP fragment to produce small, non-amyloidogenic fragments. The beta-secretase pathway involves sequential cleavages by beta-secretase and gamma-secretase complexes, and generates A-beta, which is secreted from the cell to the extracellular space, where it aggregates over time. In the central nervous system (CNS), neurons are the major sources of A-beta production. Evidence suggests that the activity of alpha-secretase, beta-secretase and gamma-secretase is dependent on CHOL metabolism. In CHO cells and various neuron-like cells grown in culture, reduction of CE either by genetic inactivation of *Acat1*, or by pharmacological inhibition of ACAT, decreases A-beta secretion (Puglielli et al. 2001). In addition, in a mouse model for AD, the ACAT inhibitors CP-113,818 and CI-1011 substantially diminished amyloid plaque density (Hutter-Paier et al. 2004; Kovacs 2006). These findings suggest that ACAT inhibitors may serve as effective chemopreventive and/or therapeutic agents for AD.

III. A working model for intracellular cholesterol trafficking

Figure 1 diagrammatically summarizes the current knowledge on trafficking of LDL-CHOL, endoCHOL, and the potential relationship between cholesterol, cholesteryl esters, and the cholesterol pool available for cellular cholesterol efflux at the PM. It shows the trafficking/recycling routes of three major cholesterol pools: cholesterol derived from low-density lipoprotein (LDL), cholesterol synthesized de novo in the endoplasmic reticulum (ER), and cholesterol involved in the cholesterol/cholesteryl ester (CE) cycle.

Cholesterol Trafficking & Esterification

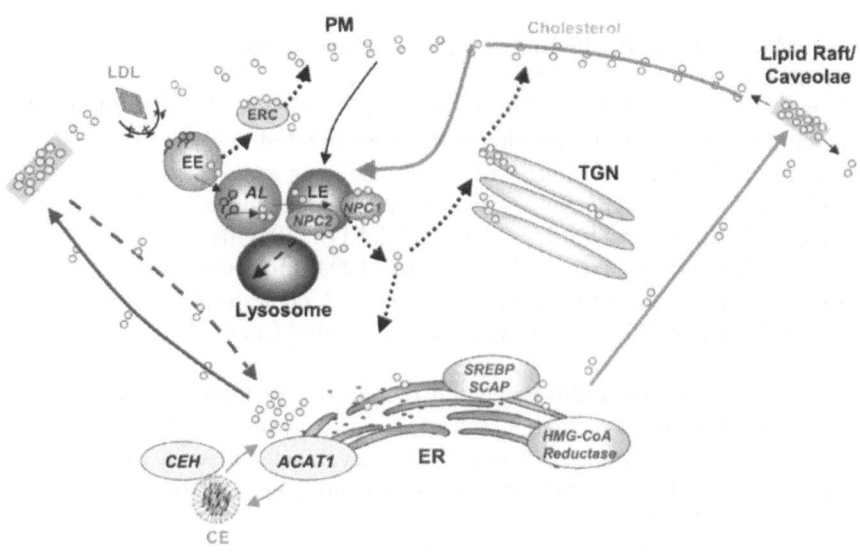

Fig. 1. Light yellow circles represent cholesterol molecules. This model is an extension and revision of the earlier model drawn in (Chang et al. 2006) . The plasma membranes (PMs) contain the highest concentration of cholesterol. The cholesterol-sensing membrane proteins are located in the ER [HMG-CoA reductase (HMGR), SREBP cleavage–activating protein (SCAP), and acyl-coenzyme A:cholesterol acyltransferase 1 (ACAT1)] or in the late endosomes [Niemann-Pick type C1 (NPC1)]. The translocation of cholesterol between various compartments may involve both vesicular and nonvesicular mechanisms. The dotted lines represent cholesterol trafficking steps that are not well documented. Other abbreviations used: AL, acid lipase; CEH, cholesteryl ester hydrolase; EE, early endosome; ERC, endocytic recycling compartment; LE, late endosome; NPC2, Niemann-Pick type C2; SREBP, sterol-regulatory element–binding protein; TGN, trans-Golgi network. Refer to color plates

Acknowledgements: The research carried out in the Chang laboratory is supported by NIH grants HL36709 and HL60306. We thank Helina Josephson for careful editing of this manuscript.

References

Brown MS, Goldstein JL (1986) A receptor-mediated pathway for cholesterol homeostasis. Science 232: 34-47

Buhman KF, Accada M, Farese RV Jr (2000) Mammalian acyl-CoA:cholesterol acyltransferases. Biochim Biophys Acta 1529: 142-154

Cadigan KM, Spillane DM, Chang TY (1990) Isolation and characterization of Chinese hamster ovary cell mutants defective in intracellular low density lipoprotein-cholesterol trafficking. J Cell Biol 110: 295-308

Chang CC, Huh HY, Cadigan KM, Chang TY (1993) Molecular cloning and functional expression of human acyl-coenzyme A:cholesterol acyltransferase cDNA in mutant Chinese hamster ovary cells. J Biol Chem 268(28): 20747-20755

Chang TY, Limanek JS (1980) Regulation of cytosolic acetoacetyl coenzyme A thiolase, 3-hydroxy-3-methylglutaryl coenzyme A synthase, 3-hydroxy-3-methylglutaryl coenzyme A reductase, and mevalonate kinase by low density lipoprotein and by 25-hydroxycholesteral in Chinese hamster ovary cells. J Biol Chem 255: 7787-7795

Chang TY, Reid PC, Sugii S, Ohgami N, Cruz JC, Chang CCY (2005) The Niemann-Pick type C disease and intracellular cholesterol trafficking. J Biol Chem 280(22): 20917-20920

Chang TY, Chang CC, Ohgami N, Yamauchi Y (2006) Cholesterol Sensing, Trafficking, and Esterification. Annu Rev Cell Dev Biol 22: 129-157

Cheruku SR, Xu Z, Dutia R, Lobel P, Storch J (2006) Mechanism of cholesterol transfer from the Niemann-Pick type C2 protein to model membranes supports a role in lysosomal cholesterol transport. J Biol Chem 281(42): 31594-31604

Choudhury A, Dominguez M, Puri V, Sharma DK, Narita K, Wheatley CL, Marks DL, Pagano RE (2002) Rab proteins mediate Golgi transport of caveola-internalized glycosphingolipids and correct lipid trafficking in Niemann-Pick C cells. J Clin Invest 109(12): 1541-1550

Cruz JC, Chang TY (2000) Fate of endogenously synthesized cholesterol in Niemann-Pick Type C1 cells. J Biol Chem 275: 41309-41316

Cruz JC, Sugii S, Yu C, Chang TY (2000) Role of Niemann-Pick type C1 protein in intracellular trafficking of low density lipoprotein-derived cholesterol. J Biol Chem 275: 4013-4021

Cuchel M, Rader DJ (2006) Macrophage reverse cholesterol transport: key to the regression of atherosclerosis? Circulation 113(21): 2548-55

Farese RV, Jr. (1998) Acyl-CoA:cholesterol acyltransferase genes and knockout mice. Curr Opin Lipidol 9: 119-124

Goldstein JL, DeBose-Boyd RA, Brown MS (2006) Protein Sensors for Membrane Sterols. Cell 124: 35-46

Guo ZY, Lin S, Heinen JA, Chang CC, Chang TY (2005) The active site His-460 of human acyl-coenzyme A:cholesterol acyltransferase 1 resides in a hitherto undisclosed transmembrane domain. J Biol Chem 280(45): 37814-26

Hao M, Lin SX, Karylowski OJ, Wustner D, McGraw TE, Maxfield FR (2002) Vesicular and non-vesicular sterol transport in living cells. The endocytic recycling compartment is a major sterol storage organelle. J Biol Chem 277(1): 609-17

Hofmann K (2000) A superfamily of membrane-bound O-acyltransferases with implications for Wnt signaling. Trends Biochem Sci 25: 111-112

Hutter-Paier B, Huttunen HJ, Puglielli L, Eckman CB, Kim DY, Hofmeister A, Moir RD, Domnitz SB, Frosch MP, Windisch M, Kovacs DM (2004) The ACAT inhibitor CP-113,818 markedly reduces amyloid pathology in a mouse model of Alzheimer's disease. Neuron 44(2): 227-38

Klansek JJ, Warner GJ, Johnson WJ, Glick JM (1996) Compartmental isolation of cholesterol participating in the cytoplasmic cholesteryl ester cycle in Chinese hamster ovary 25-RA cells. J Biol Chem 271(9): 4923-4929

Kovacs DM (2006) Oral presentation at the Keystone Symposium on Alzheimer's disease

Lange Y, Ye J, Steck TL (1998) Circulation of cholesterol between lysosomes and the plasma membrane. J Biol Chem 273: 18915-18922

Leon C, Hill JS, Wasan KM (2005) Potential role of acyl-coenzyme A:cholesterol transferase (ACAT) Inhibitors as hypolipidemic and antiatherosclerosis drugs. Pharm Res 22(10): 1578-88

Libby P, Aikawa M (2002) Stabilization of atherosclerotic plaques: new mechanisms and clinical targets. Nat Med 8: 1257-62

Lin S, Lu X, Chang CCY, Chang TY (2003) Human Acyl-Coenzyme A: Cholesterol Acyltransferase 2 (hACAT2) expressed in Chinese hamster ovary Cells: Membrane Topology and Active Site Location. Mol Biol Cell 14(6): 2447-2460

Liscum L, Sturley SL (2004) Intracellular trafficking of Niemann-Pick C proteins 1 and 2: obligate components of subcellular lipid transport. Biochim Biophys Acta 1685(1-3): 22-7

Liscum L, Ruggiero RM, Faust JR (1989) The intracellular transport of low density lipoprotein-derived cholesterol is defective in Niemann-Pick type C fibroblasts. J Cell Biol 108(5): 1625-1636

Liu J, Chang CC, Westover EJ, Covey DF, Chang TY (2005) Investigating the allosterism of Acyl Coenzyme A: cholesterol acyltransferase (ACAT) by using various sterols: In vitro and intact cell studies. Biochem J 391: 389-397

Loftus SK, Morris JA, Carstea ED, Gu JZ, Cummings C, Brown A, Ellison J, Ohno K, Rosenfeld MA, Tagle DA, Pentchev PG, Pavan WJ (1997) Murine model of Niemann-Pick C disease: mutation in a cholesterol homeostasis gene. Science 277(5323): 232-5

Matveev S, Li XL, Everson W, Smart E (2001) The role of caveolae and caveolin in vesicle-dependent and vesicle-independent trafficking. Advanced Drug Delivery Reviews 49: 237-250

Miyazaki A, Sakashita N, Lee O, Takahashi K, Horiuchi S, Hakamata H, Morganelli PM, Chang CC, Chang TY (1998) Expression of ACAT-1 protein in human atherosclerotic lesions and cultured human monocytes-macrophages. Arterioscler Thromb Vasc Biol 18(10): 1568-74

68

Naureckiene S, Sleat DE, Lackland H, Fensom A, Vanier MT, Wattiaux R, Jadot M, Lobel P (2000) Identification of HE1 as the second gene of Niemann-Pick C disease. Science 290(5500): 2298-2301

Nissen SE, Tuzcu EM, Brewer HB, Sipahi I, Nicholls SJ, Ganz P, Schoenhagen P, Waters DD, Pepine CJ, Crowe TD, Davidson MH, Deanfield JE, Wisniewski LM, Hanyok JJ, Kassalow LM (2006) Effect of ACAT inhibition on the progression of coronary atherosclerosis. N Engl J Med 354(12): 1253-63

Ohgami N, Ko DC, Thomas M, Scott MP, Chang CC, Chang TY (2004) Binding between the Niemann-Pick C1 protein and a photoactivatable cholesterol analog requires a functional sterol-sensing domain. Proc Natl Acad Sci USA 101: 12473-8

Parini P, Davis M, Lada AT, Erickson SK, Wright TL, Gustafsson U, Sahlin S, Einarsson C, Eriksson M, Angelin B, Tomoda H, Omura S, Willingham MC, Rudel LL (2004) ACAT2 is localized to hepatocytes and is the major cholesterol-esterifying enzyme in human liver. Circulation 110(14): 2017-23

Patterson MC, Vanier MT, Suzuki K, Morris JA, Carstea E, Neufeld EB, Blanchette-Mackie JE, Pentchev PG (2001) Niemann-Pick disease type C: A lipid trafficking disorder. The Metabolic and Molecular Bases of Inherited Disease. Scriver, CR, Beaudet, AL, Sly, WS and Valle, D. New York, New York, McGraw-Hill. III: 3611-3633

Pfeffer S, Aivazian D (2004) Targeting Rab GTPases to distinct membrane compartments. Nat Rev Mol Cell Biol 5(11): 886-96

Pipalia NH, Hao M, Mukherjee S, Maxfield FR (2007) Sterol, protein and lipid trafficking in Chinese hamster ovary cells with Niemann-Pick type C1 defect. Traffic 8(2): 130-41

Puglielli L, Konopka G, Pack-Chung E, MacKenzie Ingano LA, Berezovska O, Hyman BT, Chang TY, Tanzi RE, Kovacs DM (2001) Acyl-coenzyme A:cholesterol acyltransferase modulates the generation of the amyloid beta-peptide. Nat Cell Biol 3: 905-912

Raffai RL, Weisgraber KH (2003) Cholesterol: from heart attacks to Alzheimer's disease. J Lipid Res 44(8): 1423-30

Reid PC, Sugii S, Chang TY (2003) Trafficking defects in endogenously synthesized cholesterol in fibroblasts, macrophages, hepatocytes, and glial cells from Niemann-Pick type C1 mice. J Lipid Res 44: 1010-1019

Sleat DE, Wiseman JA, El-Banna M, Price SM, Verot L, Shen MM, Tint GS, Vanier MT, Walkley SU, Lobel P (2004) Genetic evidence for nonredundant functional cooperativity between NPC1 and NPC2 in lipid transport. Proc Natl Acad Sci USA 101: 5886-5891

Song BL, Wang CH, Yao XM, Yang L, Zhang WJ, Wang ZZ, Zhao XN, Yang JB, Qi W, Yang XY, Inoue K, Lin ZX, Zhang HZ, Kodama T, Chang CC, Liu YK, Chang TY, Li BL (2006) Human acyl-CoA:cholesterol acyltransferase 2 gene expression in intestinal Caco-2 cells and in hepatocellular carcinoma. Biochem J 394(Pt 3): 617-26

Sugii S, Reid PC, Ohgami N, Du H, Chang TY (2003) Distinct endosomal compartments in early trafficking of low density lipoprotein-derived cholesterol. J Biol Chem 278(29): 27180-27189

Sugii S, Lin S, Ohgami N, Ohashi M, Chang CC, Chang TY (2006) Roles of endogenously synthesized sterols in the endocytic pathway. J Biol Chem 281(32): 23191-206

Walter M, Davies JP, Ioannou YA (2003) Telomerase immortalization upregulates Rab9 expression and restores LDL cholesterol egress from Niemann-Pick C1 late endosomes. J Lipid Res 44: 243-253

Wojtanik KM, Liscum L (2003) The transport of LDL-derived cholesterol to the plasma membrane is defective in NPC1 cells. J Biol Chem 278(17): 14850-14856

Wolozin B (2004) Cholesterol, statins and dementia. Curr Opin Lipidol 15: 667-72

Ye S, Huang Y, Mullendorff K, Dong L, Giedt G, Meng EC, Cohen FE, Kuntz ID, Weisgraber KH, Mahley RW (2005) Apolipoprotein (apo) E4 enhances amyloid beta peptide production in cultured neuronal cells: apoE structure as a potential therapeutic target. Proc Natl Acad Sci U S A 102(51): 18700-5

Yu C, Zhang Y, Lu X, Chang CCY, Chang TY (2002) The role of the N-terminal hydrophilic domain of acyl coenzyme A:cholesterol acyltransferase 1 on the enzyme's quaternary structure and catalytic efficiency. Biochemistry 41: 3762-3769

Yu C, Chen J, Lin S, Liu J, Chang CCY, Chang TY (1999) Human acyl-CoA:cholesterol acyltransferase-1 is a homotetrameric enzyme in intact cells and in Vitro. J Biol Chem 274: 36139-36145

Atherogenic Lipoproteins in Type 2 Diabetes with Nephropathy

Tsutomu Hirano
Division of Diabetes and Metabolism, First Department of Internal Medicine, Showa University School of Medicine, 1-5-8 Hatanodai, Shinagawa-ku, Tokyo 142-8666, Japan.

Summary. It is well known patients with diabetes have a high incidence of cardiovascular disease (CVD), and the incidence of CVD becomes substantially elevated when diabetic nephropathy is developed. The pathogenesis of CVD in diabetes is multifactorial, but dyslipidemia is thought to be a powerful risk factor for CVD. Although lipid metabolism has been extensively investigated in diabetes, little information is available concerning the influence of nephropathy on diabetic dyslipidemia. The mechanisms for dyslipidemia in diabetic nephropathy are multifactrial and complex. Plasma lipid profile are largely changed in the stage of nephropathy. Diabetes per.se. is a basic component causing plasma lipid abnormalities. Long-term hyperglycemia causes generalized vascular endothelial damage, which reduces functional lipoprotein lipase, leading to increase TG and decrease HDL-cholesterol. In overt-diabetic nephropathy, hypoproteinemia markedly increases LDL-C, and renal failure specifically increases remnant lipoproteins and decreases HDL-C and LDL-C. In this brief review, I will state how lipoprotein abnormalities worsen with increasing severity of diabetic nephropathy, and what lipoprotein changes are attributable to diabetes or kidney dysfunction based on our data.

Key words. Lipoprotein, Remnants, Diabetic nephropathy, CKD, CHD

1 Introduction

Patients with diabetes are well known to have a high incidence of cardio-vascular disease (CVD), and the incidence of CVD rises substantially when diabetic nephropathy develops. The pathogenesis of CVD in diabetes is multifactorial, but dyslipidemia is thought to be a powerful risk factor. Although lipid metabolism has been extensively investigated in diabetes, little information is available on the influences of nephropathy on diabetic dyslipidemia. I would like to demonstrate two things: first, how lipoprotein abnormalities worsen with increasing severity of diabetic nephropathy; second, what lipoprotein changes are attributable to diabetes or kidney dysfunction.

2 CKD and CHD

The prevalence of CHD events was significantly higher in subjects with a glomerular filtration rate (GFR) of less than 60—in other words, in the patients with what we now call chronic kidney disease (CKD) (Muntner P et al., 2005). An accumulating body of evidence has revealed that microalbuminuria is associated with a high incidence of cardiovascular disease. According to the WHO criteria, microalbuminuria is a component of the metabolic syndrome. CHD events are markedly elevated in the presence of microalbuminuria in a hyperinsulinemic state (Kuusisto J et al., 1995). It thus appears that insulin resistance and microalbuminuria synergistically promote the progression of atherosclerosis.

3 Dyslipidemia in Diabetic Nephropathy

Cardiovascular events seem to be much more frequent in patients with dia-betic nephropathy than in those without nephropathy. Our group measured the intima-media thickness of the cervical artery, a widely used surrogate marker for premature atherosclerosis. The max IMT increased with the progression of diabetic nephropathy. The mechanisms behind dyslipidemia in diabetic nephropathy are multifactorial and complex. "**Fig.1**" shows possible components involved. Diabetes in itself is a component cause of plasma lipid abnormalities. Long-term hyperglycemia leads to generalized vascular endothelial damage, which then leads to dyslipidemia. Nephrosis and renal failure are also well established as component causes of plasma lipid abnormalities. Plasma lipid profiles change substantially as the neph-

ropathy advances. In the early stage of diabetic nephropathy, hypertriglyceridemia develops. In overt-diabetic nephropathy, nephrosis markedly increases LDL-C, and renal failure increases remnant lipoprotein and
decreases HDL-C ("**Fig.2**"). Even though the fasting hypertriglyceridemic
subjects were excluded, the diabetic nephropathy group exhibited a remarkable postprandial hypetriglyceridemia in the fat-tolerance test (Hirano
T et al., 1998).

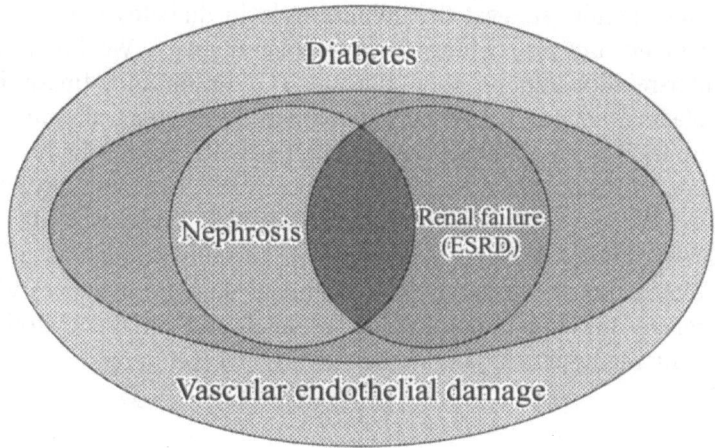

Fig.1

Fig.2

4 Dyslipidemia in Early Diabetic Nephropathy

Our group studied a possible mechanism behind the hypertriglyceridemia in early diabetic nephropathy with microalbuminuria. Plasma von Willebrand factor is widely used as a surrogated marker for vascular endothelial damage. VWF was significantly elevated in microalbuminuric and macroproteinuric diabetes, whereas it remained normal in non-diabetic patients with primary kidney disease having massive proteinuria (Hirano T et al., 2000). These results suggest that albuminuria in diabetes implies not only kidney damage, but also widespread vascular damage. We found a meaningful inter-relationship between plasma TG, lipoprotein lipase in post-heparin plasma, and vWF. When we compared the microalbuminuric diabetics with their normoalbuminuric counterparts, the former had higher TG, lower heparin-releasable LPL, and higher vWF. Interestingly, heparin-releasable LPL was inversely correlated with vWF. The lower LPL was associated with higher TG and lower HDL-C (Kashiwazaki K et al., 1998). We speculated that the generalized endothelial damage increases vWF and this decreases the functional LPL mass anchored on the endothelium , thereby inducing hypertriglyceridemia and low HDL-C.

5 Dyslipidemia in Nephrotic Syndrome

It is well known that LDL cholesterol is markedly increased in nephrotic syndrome. But the precise mechanisms behind this remain poorly understood. Several kinetic studies have demonstrated substantially increased rates of VLDL and LDL production in patients with nephrotic syndrome. In severe nephrosis, VLDL catabolism is impaired which increases TG (Yoshino G et al., 1993). HDL-C, on the other hand, is often decreased because of the higher activity of CETP. What mechanism leads to the increases in VLDL and LDL production? Yet other investigators have reported that the mRNA of apo B in the liver is not increased in nephrotic syndrome, suggesting that the posttranscriptional process is involved in the hypersecretion of apo-B-containing lipoproteins. We firstly demonstrated that intracellular apo B degradation was decreased in hepatic cells when the cells were cultured with lower concentrations of bovine serum albumin (Hirano T et al., 1995). Reduction in apo B degradation leads to increase apo B secretion into the medium. In contrast, apo B production was unchanged at different BSA concentrations. These results suggest that decreased intercellular degradation of apo B is a major cause of the overproduction of apo-B-containing lipoproteins in nephrotic syndrome.

6 Dyslipidemia in End-stage Renal Disease (ESRD)

Apo CIII is a key inhibitor of the lipolysis and particle uptake of TG-rich lipoproteins. Apo B48 is a structural protein of intestine-derived lipoproteins such a chylomicron and its remnants. Patients with renal failure undergo elevations in the production of apo CIII in the liver which eventually impair the catabolism of chylomicron remnants.

We recently found that apo B48, a marker of chylomicron remnants, increased as diabetic nephropathy progressed (Hayashi T et al., 2007). The ratio of apo B48 to TG was also substantially increased in diabetic renal failure, suggesting specific increases of chylomicron remnants in ESRD. Chylomicron remnants are atherogenic lipoprotein particles capable of penetrating the arterial wall and lodging in the sub-endothelial spaces. IDL, a major part of VLDL remnant, was also significantly increased in clinical diabetic nephropathy and in end-stage diabetic nephropathy (Hayashi T et al., 2007). Corresponding to these changes, VLDL-apo CIII levels were significantly elevated in clinical and end-stage diabetic nephropathy. VLDL-apo CIII was also higher in non-diabetic ESRD (Hirano T et al., 2003). This suggests that renal failure, not diabetes, is the main cause of the increased VLDL-apo CIII. Increased apo CIII suppresses LPL activity, which leads to hypertriglyceridemia in ESRD. HTGL, a key enzyme for hydrolysis of IDL, is markedly decreased in ESRD. HTGL is equally decreased in diabetic and non-diabetic ESRD, which may explain substantial increase of remnant lipoproteins in ESRD, irrespective of the presence of diabetes (Oi K et al., 1999). Recent kinetic studies have revealed that the catabolism of VLDL, IDL, and LDL are all impaired in ESRD (Ikewaki K et al., 2005). The long residence time of these lipoproteins is likely to increase their atherogenicity.

7 Conclusion

The atherogenic lipoprotein profile becomes more prominent in type 2 diabetics when diabetic nephropathy advances. In diabetes with ESRD, increased apo CIII and decreased HTGL are major causative factors behind the accumulation of atherogenic remnant particles in the blood circulation. Dyslipidemia associated with diabetic nephropathy should be more closely watched as a very strong risk factor for cardiovascular disease.

References

Hayashi T, Hirano T, Taira T, Tokuno A, Mori Y, Koba S, Adachi M (2007) Remarkable increase of apolipoprotein B48 level in diabetic patients with endstage renal disease. Atherosclerosis ;24 (In press)

Hirano T, Furukawa S, Kurokawa M, Ebara T, Dixon JL, Nagano S (1995) Intracellular apoprotein B degradation is suppressed by decreased albumin concentration in Hep G2 cells.Kidney Int 47:421-431.

Hirano T, Oi K, Sakai S, Kashiwazaki K, Adachi M, Yoshino G (1998) High prevalence of small dense LDL in diabetic nephropathy is not directly associated with kidney damage: a possible role of postprandial lipemia. Atherosclerosis 141:77-85.

Hirano T, Ookubo K, Kashiwazaki K, Tajima H, Yoshino G, Adachi M (2000) Vascular endothelial markers, von Willebrand factor and thrombomodulin index, are specifically elevated in type 2 diabetic patients with nephropathy: comparison of primary renal disease. Clin Chim Acta 299:65-75.

Hirano T, Sakaue T, Misaki A, Murayama S, Takahashi T, Okada K, Takeuchi H, Yoshino G, Adachi M (2003) Very low-density lipoprotein-apoprotein CI is increased in diabetic nephropathy: comparison with apoprotein CIII. Kidney Int 63:2171-2177.

Ikewaki K, Schaefer JR, Frischmann ME, Okubo K, Hosoya T, Mochizuki S, Dieplinger B, Trenkwalder E, Schweer H, Kronenberg F, Koenig P, Dieplinger H (2005) Delayed in vivo catabolism of intermediate-density lipoprotein and low-density lipoprotein in hemodialysis patients as potential cause of premature atherosclerosis. Arterioscler Thromb Vasc Biol 25:2615-2622.

Kashiwazaki K, Hirano T, Yoshino G, Kurokawa M, Tajima H, Adachi M (1998) Decreased release of lipoprotein lipase is associated with vascular endothelial damage in NIDDM patients with microalbuminuria. Diabetes Care 21:2016-2020.

Kuusisto J, Mykkanen L, Pyorala K, Laakso M (1995) Hyperinsulinemic microalbuminuria. A new risk indicator for coronary heart disease. Circulation. 91:831-837.

Muntner P, He J, Astor BC, Folsom AR, Coresh J (2005) Traditional and nontraditional risk factors predict coronary heart disease in chronic kidney disease: results from the atherosclerosis risk in communities study. J Am Soc Nephrol 16:529-538.

Oi K, Hirano T, Sakai S, Kawaguchi Y, Hosoya T (1999) Role of hepatic lipase in intermediate-density lipoprotein and small, dense low-density lipoprotein formation in hemodialysis patients. Kidney Int 71:S227-S228.

Yoshino G, Hirano T, Nagata K, Maeda E, Naka Y, Murata Y, Kazumi T, Kasuga M (1993) Hypertriglyceridemia in nephrotic rats is due to a clearance defect of plasma triglyceride: overproduction of triglyceride-rich lipoprotein is not an obligatory factor. J Lipid Res 34:875-884.

Regulatory mechanisms for cytosolic prostaglandin E synthase, cPGES/p23

Yoshihito Nakatani and Ichiro Kudo

Department of Health Chemistry, School of Pharmaceutical Sciences, Showa University, 1-5-8 Hatanodai, Shinagawa-ku, Tokyo 142-8555, Japan
<e-mail> nakatani@pharm.showa-u.ac.jp

Summary. We reported the molecular identification of cytosolic prostaglandin E synthase (cPGES), a terminal enzyme of the cyclooxygenase-mediated PGE_2 biosynthetic pathway. Of interest, it is identical to the co-chaperone p23 that binds to heat shock protein 90 (Hsp90). Association of cPGES/p23 and Hsp90 resulted in a remarkable increase in PGES activity *in vitro*. Next, we found that cPGES/p23 underwent serine phosphorylation, which was accelerated transiently after cell activation. In activated cells, cPGES/p23 phosphorylation occurred in parallel with increased cPGES/p23 enzymic activity and PGE_2 production from exogenous and endogenous arachidonic acid, and these processes were facilitated by Hsp90 that formed a tertiary complex with cPGES/p23 and protein kinase CK2. Treatment of cells with inhibitors of CK2 and Hsp90 and with a dominant-negative CK2 attenuated the formation of the cPGES/p23-CK2-Hsp90 complex and attendant cPGES/p23 phosphorylation and activation. Mutations of either of two predicted CK2 phosphorylation sites on cPGES/p23 (Ser^{113} and Ser^{118}) abrogated its phosphorylation and activation both *in vitro* and *in vivo*. These results provide the evidence that the cellular function of this eicosanoid-biosynthetic enzyme is under the control of a molecular chaperone and its client protein kinase.

Keywords: prostaglandin E synthase, molecular chaperone, phosphorylation, prostaglandin E_2, protein kinase CK2, heat shock protein 90

1 Introduction

Cytosolic prostaglandin E synthase (cPGES) is a 23-kDa glutathione (GSH)-requiring enzyme expressed constitutively in a wide variety of cells. This enzyme is selectively coupled with COX-1 to mediate immediate PGE_2 production. Of particular note, cPGES is identical to the co-chaperone p23, which is associated with heat shock protein 90 (Hsp90) (Tanioka et al. 2000). In contrast, the GSH-dependent, membrane-bound PGES (mPGES-1) is markedly induced by proinflammatory stimuli, down-regulated by glucocorticoids, and functionally coupled with COX-2 (Murakami et al. 2000). mPGES-2 has a glutaredoxin- or thioredoxin-like domain and is activated by several thiol reagents. This enzyme is constitutively expressed in various cells and can be coupled with both COX-1 and COX-2 (Murakami et al. 2003).

Here, we briefly review our recent studies on the regulatory mechanism of cPGES/p23 in rat fibroblastic 3Y1 cells. Association of cPGES/p23 with Hsp90 was increased in cells stimulated with Ca^{2+} mobilizer, accompanied by concomitant increases in cPGES/p23 enzymic activity and PGE_2 production (Tanioka et al. 2003). cPGES/p23 also underwent phosphorylation by protein kinase CK2, leading to increase in cPGES/p23 enzymic activity and association of cPGES/p23 with Hsp90 (Kobayashi et al. 2004). Taken together, our results suggest that the cellular PGE_2-biosynthetic function is under the control of a molecular chaperone and its client protein kinase.

2 Regulation by Hsp90

Incubation of bacterial expression system-derived recombinant cPGES/p23 with Hsp90 *in vitro* in the presence of ATP and Mg^{2+} resulted in not only the formation of cPGES/p23-Hsp90 complex, but also a remarkable increase in cPGES activity. This result is consistent with the fact that both ATP and Mg^{2+} are necessary for the formation of the cPGES/p23-Hsp90 complex. Next, we examined interaction between cPGES/p23 and Hsp90 in rat fibroblastic 3Y1 cells, in which cPGES/p23 is the predominant PGES enzyme. Stimulation of 3Y1 cells with Ca^{2+}-ionophore A23187 elicited immediate PGE_2 synthesis. The increase of PGE_2 synthesis was preceded by that of cPGES/p23 enzymic activity in cell lysates, which occurred within 1 min and peaked at 15 min. Immunoprecipitation of the cell lysates with anti-cPGES/p23 antibody revealed that more Hsp90 was co-precipitated with cPGES/p23 from A23187-treated cells than from untreated cells, indicating that there was an increased association between cPGES/p23 and Hsp90 following cell activation. To assess whether cPGES/p23 activation occurs even by physiological stimulus, 3Y1 cells were treated with bradykinin. Stimulation of cells with bradykinin resulted in increased PGE_2 synthesis and PGES activity in cell lysates, which were accompanied by increased association between cPGES/p23 and Hsp90.

Treatment of the cells with the Hsp90 inhibitors, geldanamycin and novobiocin, each of which binds to the ATP-binding site on Hsp90 and inhibits its chaperone function, reduced this stimulus-increased association between cPGES/p23 and Hsp90 to a basal level. The dissociation of the cPGES/p23-Hsp90 complex by the Hsp90 inhibitors was accompanied by suppression of bradykinin-induced PGE_2 synthesis and cPGES/p23 activation. Thus, it appears that stimulus-induced formation of the cPGES/p23-Hsp90 complex is correlated with the activity of cPGES/p23 to synthesize PGE_2 in cells.

3 Regulation by phosphorylation

There are several potential phosphorylation sites in the primary amino acid sequence of human cPGES/p23. Judging from the effects of several protein kinase inhibitors on conversion of exogenous arachidonic acid to PGE_2 in 3Y1 cells, protein kinase CK2 may be involved in phosphorylation and activation of cPGES/p23, and protein kinase C (PKC) and Ca^{2+}/calmodulin kinase II (CaMKII) are not. To verify the contribution of CK2 to cPGES/p23 activation, we took advantage of DN-CK2, which harbors a mutation (D156A) in the kinase domain and thereby blocks the cellular functions of CK2. Overexpression of DN-CK2 in 3Y1 cells resulted in decrease of phosphorylation of cPGES/p23 and association of cPGES/p23 with Hsp90 relative to those in mock-transfected cells. These results indicate that complex formation of cPGES/p23 with CK2 and Hsp90, and concomitant phosphorylation by CK2 are essential for the activation of cPGES/p23 in cells.

To address which residues of cPGES/p23 are phosphorylated by CK2 in cells, we constructed point mutants of cPGES/p23 in which each CK2-consensus Ser residue was replaced with Ala or Gly (S113A, S118A and S151G). PGE_2 synthesis induced by bradykinin was markedly elevated when WT or S151G was transfected, whereas in the case of S113A or S118A it was not significantly altered. Immunoprecipitation analysis of ^{32}P-pre-labeled cells revealed that phosphorylation of cPGES/p23 occurred in WT- or S151G-transfected cells, and not in S113A or S118A-transfected cells. Judging from these results, we conclude that Ser[113] and

Ser[118] are the predominant CK2 phosphorylation sites critical for cPGES/p23 activation in cells, and that phosphorylation of these two Ser residues promotes association of cPGES/p23 with Hsp90.

Conclusion

We propose that functional cPGES/p23 exists in cells as a multicomponent complex containing CK2 and Hsp90 as the minimal requirements. CK2 activated by upstream signals triggers dual phosphorylation of Ser[113] and Ser[118] on cPGES/p23, which in turn facilitates the recruitment of cPGES/p23 to the Hsp90 complex, leading to its full activation. In porcine endothelial cells, bradykinin induces phosphorylation of angiotensin-converting enzyme by CK2, consistent with our findings that bradykinin elicits CK2-dependent cPGES/p23 activation, although the signaling from bradykinin to CK2 activation remains to be elucidated. Our results have provided the evidence that the function of PGES is elegantly controlled by a particular protein kinase in cooperation with a molecular chaperone.

References

Kobayashi T, Nakatani Y, Tanioka T, Tsujimoto M, Nakajo S, Nakaya K, Murakami M and Kudo I. (2004) Regulation of cytosolic prostaglandin E synthase by phosphorylation. *Biochem J* **381** 59-69

Murakami M, Naraba H, Tanioka T, Semmyo N, Nakatani Y, Kojima, F, Ikeda T, Fueki M, Ueno A, Oh-ishi S and Kudo I. (2000) Regulation of prostaglandin E_2 biosynthesis by inducible membrane-associated prostaglandin E_2 synthesis that acts in concert with cyclooxygenase-2. *J Biol Chem* **275** 327783-32792

Murakami M, Nakashima K, Kamei D, Masuda S, Ishikawa Y, Ishii T, Ohmiya Y, Watanabe K Kudo I (2003) Cellular prostaglandin E_2 production by membrane-bound prostaglandin E synthase-2 via both cyclooxygenases-1 and -2. *J Biol Chem* **278** 37937-37947

Tanioka T, Nakatani Y, Semmyo N, Murakami M and Kudo I. (2000) Molecular identification of cytosolic prostaglandin E_2 synthase that is functionally coupled with cyclooxygenase-1 in immediate prostaglandin E_2 biosynthesis. *J Biol Chem* **275** 32775-32782

Tanioka T, Nakatani Y, Kobayashi T, Tsujimoto M, Oh-ishi S, Murakami M and Kudo I. (2003) Regulation of cytosolic prostaglandin E_2 synthase by 90-kDa heat shock protein. *Biochem Biophys Res Commun* **303** 1018-1023

Regulation of Intracellular Lipid Storage and Adipose Differentiation-Related Protein (ADRP)

Hiroyuki Itabe[1,2], Yutaka Masuda[1], Naoko Sasabe[1], Keiko Kitazato[1], Hiroyuki Arai[3] and Tatsuya Takano[2]

[1]Department of Biological Chemistry, School of Pharmaceutical Sciences, Showa University, 1-5-8 Hatanodai, Shinagawa-ku, Tokyo 142-8555, Japan <e-mail> h-itabe@pharm.showa-u.ac.jp
[2]Department of Molecular Pathology, Faculty of Pharmaceutical Sciences, Teikyo University, 1091-1 Suarashi, Sagamihara, Kanagawa 199-0195, Japan
[3]Department of Health Chemistry, Graduate School of Pharmaceutical Sciences, Tokyo University, 7-3-1 Hongo, Bunkyo-ku, Tokyo, 113-0033

Summary. Adipose differentiation-related protein (ADRP) is a major protein localized at the lipid droplets in macrophage-derived foam cells or liver cells. However, the role of ADRP during regression of lipid-storing cells has not been understood. When J774 macrophages were incubated for 3 days with VLDL, their content of ADRP and triacylglycerol (TG) increased 3- and 4-fold, respectively. Induction of ADRP was observed by loading lipids in macrophages with either oleic acid, suggesting that the induction ADRP induction was not receptor-dependent. ADRP expression during TG accumulation was also induced in oleic acid-treated HuH-7 human liver cells. As TG decreased in the foam cells by treatment with triacsin C, an acyl-CoA synthase inhibitor, for 6 h, ADRP protein decreased in parallel. Treatment of lipid-stored HuH-7 cells with triacsin C also reduced ADRP, indicating that ADRP reduced during regression of the lipid-storing cells. Such decrease in ADRP during regression of TG-storing cells was abolished by co-incubation with a proteasome inhibitor. Poly-ubiquitinated ADRP was detected by a pull-down experiment in the presence of the proteasome inhibitor. In addition, the proteasome inhibitor reversed not only the degradation of ADRP but TG loss by triacsin C in HuH-7 cells. The amount of ADRP is reciprocally regulated with the lipid content in the cells, and the ubiquitin-proteasome system is involved in degradation of ADRP during regression of lipid-storing cells.

Key words. Lipd droplets, adipose differentiation related protein, macrophages, hepatocytes, proteasome.

Introduction

Recently attention has been paid on cytosolic lipid droplets (LD), since they can be important cellular machinery for lipid metabolism in physiological and/or pathological conditions (Londos et al. 1999, Brown 2001). Not only adipocytes, the major lipid-storing center of the body, contain full of LD, but foam cells found in atherosclerotic lesions and parenchymal cells in fatty liver are also LD-containing cells. They accumulate massive amount of cholesteryl ester (CE) and/or triacylglyerol (TG) as LD in their cytoplasmic space. It is speculated that the LD is an organized intracellular structure like other organelles.

Adipose differentiation-related protein (ADRP, also called as ADFP or adipophilin), is a major LD-associating protein in the liver and macrophages (Fujimoto et al. 2004) (Fig. 1). ADRP was originally found as a protein markedly induced in early stages of adipocyte differentiation, although ADRP on LD is replaced by another LD-associated protein, perilipin A, in mature adipocytes (Brasaemle et al. 1997).

ADRP and perilipin A localize on the surface of intracellular lipid droplets, and they share a homologous sequence in the N-teminal regions called PAT domain (Fig. 2) (Londos et al. 1999). In mature adipocytes, perilipin A stabilizes the droplets structure and also regulates lipolysis of TG in the LD by hormone-sensitive lipase. Upon stimulation with glucagon or adrenalin, perilipin A is phosphorylated by protein kinase A, then the hormone-sensitive lipase associates with LD through binding to

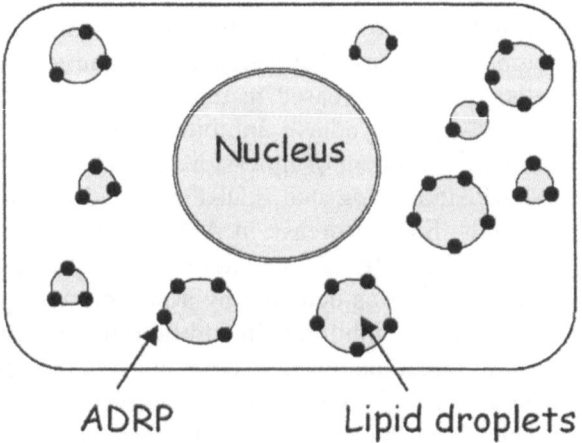

Fig. 1 ADRP is localized on the surface of intracellular LD in various cells including macrophages and hepatocytes.

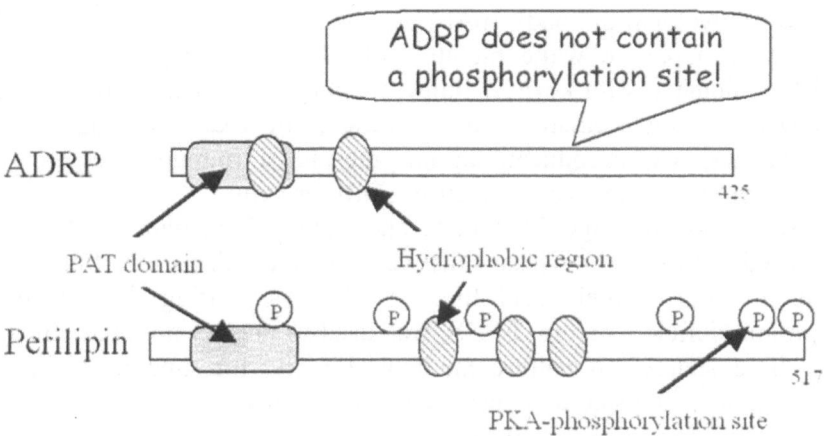

Fig. 2. Comparison of ADRP and perilipin A, two major LD-associated proteins.

the phoshorylated form of perilipin A and reacts with its substrates within the LD. Perilipin-null mice showed extensive reduction in their adipose tissue mass because of elevated lipolysis (Birnbaum et al. 2003). ADRP does not have any putative phosphorylation site judging by the amino acid sequence, which is a different feature from perilipin A. Thus, it is speculated that lipid metabolism in ADRP-coated LD may be regulated by other regulatory mechanisms, although ADRP and perilipin A share similar properties for LD formation.

It is well known that foam cells can be regressed under certain conditions. Cellular lipid storage is largely controlled by the balance between influx and efflux of lipids, and regression of lipid-storing cells should be a way of therapeutic approaches. In this study we questioned what happens during regression of lipid droplets (Masuda et al. 2006). To approach this issue, we quantified the amount of ADRP protein expressed in macrophages and in liver cells using anti-ADRP antibodies together with the amount of TG. We found that ADRP decreases concomitant with cellular TG reduction and that the ubiquitin-proteasome system is involved in ADRP reduction during regression of lipid-accumulating cells.

Results

ADRP protein expression during lipid loading

A rabbit polyclonal antibody (pAb) against mouse ADRP was raised by immunizing Japanese white rabbits with C-terminal peptide of murine ADRP conjugated with keyhole limpet hemocyanin. Quantitative changes of ADRP during formation and regression of foam cells in murine macrophages, we introduced an enzyme-linked immunosorbent assay (ELISA) procedure to measure mouse ADRP protein.

Mouse macrophage J774 cells and human hepatoma HuH-7 cells were used in this study. These cells were treated with various lipids such as lipoproteins or oleic acid for 1 or 3 days. During the treatment, lipoprotein deficient serum, LPDS, was used to neglect possible effect of other lipid sources in culture medium. Finally, the amount of ADRP protein in the cell lysate was measured using antibody.

Macrophages were incubated with the same amounts (50 µg/ml) of LDL, VLDL or acetylated LDL for 3 days. Under the conditions, VLDL was the most effective for the increase in the amount of cellular lipids. The cellular TG increased 3-fold. ADRP protein was also increased 3-fold by the treatment with VLDL.

Since VLDL treatment increased ADRP protein, we measured mRNA levels. For this purpose, macrophages were incubated with one of those lipoproteins (50 µg/ml) for 2 day, and the amount of ADRP mRNA was evaluated by real-time PCR method. ADRP mRNA was induced as much as 5-fold by the treatment with VLDL.

Human hepatic cell line, HuH-7 were incubated with 600 µM of oleic acid conjugated with BSA for up to 24 h. Both cellular TG and ADRP protein increased 10 and 6 times under the conditions. HuH-7 cells contain number of small LD before lipid loading, and the LD increase their size as well as the number by treatment with oleic acid. We thought that concomitant increase in TG and ADRP protein was not specific event in macrophage-derived foam cells. Presumably it may be a common response in many cells.

ADRP protein reduced during regression of lipid-storing cells

As described above, when cells were treated with lipoprotein or free fatty acid, they accumulate TG and the amount of ADRP protein increased. It is very likely the induction of ADRP and formation of lipid droplets are closely coupled. Then, we questioned what happens on ADRP protein when these cells reduce their intracellular TG and regress their LD.

HuH-7 cells were incubated with oleic acid for the 24 h, then further incubated without oleic acid but with triacsin C, a potent inhibitor for acyl CoA-synthetase, for up to 24 h. By such treatment, the amount of TG increased by the first 24 h loading, and it decreased by 50 % during the following incubation with triacsin C. ADRP increased by treatment with oleic acid for 24 h, and it also decreased by 60 % during the 2nd incubation under the conditions. During regression of macrophage foam cells, a similar reduction of ADRP was also observed.

Such a rapid reduction of cellular ADRP protein might be caused by active degradation of protein. We examined a possibility that proteasome is involved in the reduction of ADRP. HuH-7 cells were incubated with oleic acid for the first 24 h, then further incubated without oleic acid for 15 h. This time, a proteasome inhibitor, MG132, in addition to triacsin C was added to the medium during the 2nd incubation (Fig. 3). In the absence of MG132, the amounts of TG and ADRP decreased by 50 % during the 2nd incubation. However, addition of MG132 blocked the reduction of ADRP in a dose-dependent manner. More importantly, MG132 also blocked the

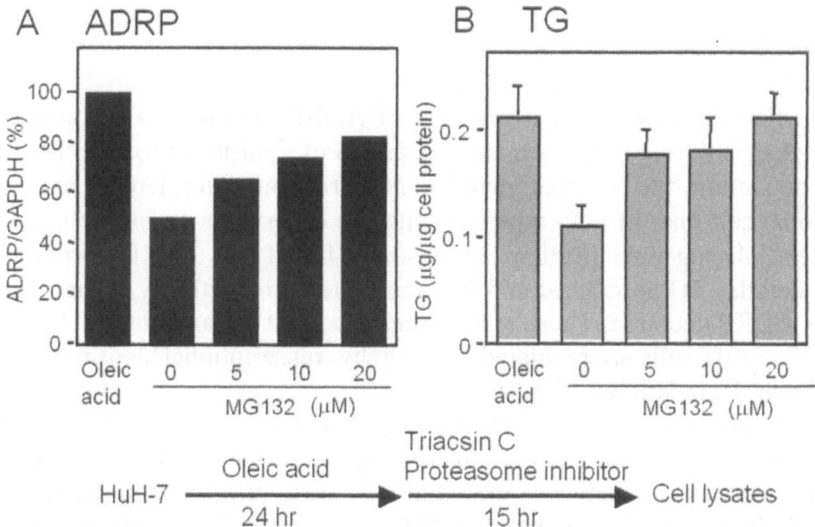

Fig. 3. The proteasome inhibitor abolishes the decrease in ADRP protein during regression of oleic acidtreated HuH-7cells. HuH-7 cells (1×10^6 cells/dish) were incubated for 24 h in medium containing 600 µM oleic acid. Then the cells were further incubated for 15 h in new medium with 5% LPDS, 25 mg/ml BSA, 10 µg/ml triacsin C, and various concentrations of MG132. ADRP and GADPH in the cell lysates were determined by immunoblotting. The intensity of the bands was quantified with NIH Image, and the amount of ADRP relative to GAPDH was calculated (A). Under the same experimental conditions, the amounts of TG in the cell lysates were measured by the enzymatic method (B). Shown are typical data in three independent experiments. (cited from Masuda, et al. J Lipid Res 47 (2006) 87-98 with a permission.)

reduction of TG under the same condition, suggesting that ADRP could function as a stabilizer of LD by preventing TG from hydrolysis.

Prevention of ADRP reduction by MG132 strongly suggests that ADRP is degraded by ubiquitin-proteasome pathway. The presence of ubiquitinated ADRP was proved by an immunoprecipitation experiment. To precipitate ADRP effectively, a DNA construct for ADRP tagged with FLAG sequence was transfected into HuH-7 cells. The transfected cells were incubated with the proteasome inhibitor MG132 but no lipid sources. The FLAG-tagged protein was recovered using anti-FLAG antibody, then ubiquitin and ADRP in the precipitant were detected by western blotting. There are smear bands detected with anti-ubiquitin antibody with molecular weight higher than ADRP, but no bands were observed without MG132 treatment. These results clearly demonstrate that ADRP is ubiquitinated under lipid-poor conditions.

Discussion

We studied the changes in expression of ADRP protein in J774 murine macrophages during formation and regression of lipid-laden foam cells. In addition similar sets of experiments were carried out using HuH-7 human hepatome cell line. In both types of cells, we observed a strong induction of ADRP during lipid loading, and also we found that ADRP decreased when cellular TG is reduced and that the proteasome pathway is involved in this ADRP decrease. These results suggested that the amounts of ADRP in lipid-storing cells are regulated not only by transcriptional steps but also by protein-degradation systems.

Fig. 4 illustrates our current working hypothesis. Addition of lipid sources to cells, such as lipoproteins or fatty acids, induces LD formation. ADRP protein is induced concomitant with the accumulation of intracellular lipids, presumably ADRP is needed for LD formation. When the lipid storing cells lose their lipids, ADRP protein reduces concomitant with the reduction of lipids. This decrease in ADRP was regulated by ubiquitin-proteasome pathway. ADRP is likely to have a function in stability of LD since inhibition of ADRP degradation prevented lipid loss. From our observations, we would like to propose that ADRP is an important factor for intracellular lipid metabolism, and its regulatory mechanism is different from that of perilipin.

It was shown that ADRP is expressed through activation of PPARδ in macrophages, where peritoneal macrophages obtained from PPARδ-null mice was not able to induce ADRP (Chawla et al. 2003). They showed close relationship between TG accumulation and PPARδ activation by reporter-gene assay. Activation of PPARδ by a selective agonist promoted lipid accumulation in human primary macrophages (Larigauderie et al. 2004). The physiological ligand(s) for PPARδ has yet to be clarified, however, fatty acids rather than TG are probable candidates. In our study, the finding that the change in amount of TG corresponds to the change in ADRP more than those of cholesterol is well accordance with the possible involvement of PPARδ.

The physiological role of ADRP has yet to be fully understood. Recently it was reported that ADRP-knockout mice did have intracellular LD (Chang 2006). Although the size of LD recovered from the liver was significantly smaller in ADRP-knockout mice than in wild-type mice when the mice were treated with high fat diet. LD was present even in the absence of ADRP, since ADRP can be replaced by TIP-47, another LD-associated proteins. Such redundancy would suggest the physiological importance of LD, however, it might be difficult to elucidate physiological role of ADRP by gene-knockout approaches. Our study focused on actual amount of ADRP protein rather than gene expression levels could be a reasonable way to study the role of ADRP. Indeed we found an evidence to show that ADRP could stabilize LD and protect the lipids in LD from the hydrolytic activity (Fig. 3).

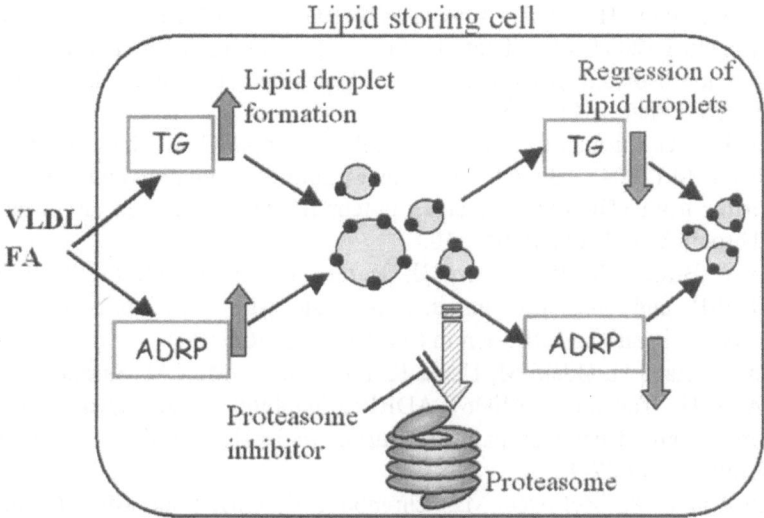

Fig. 4. Current hypothesis on regulatory mechanisms of ADRP and lipid droplets.

In conclusion, we found that ubiquitin pathway is involved in regulation of ADRP during regression of macrophage-derived foam cells and lipid-laden liver cells. This may be one type of regulatory features of hormone-independent components of lipid-associating proteins.

References

Birnbaum MJ (2003) Lipolysis: more than just a lipase. J Cell Biol 161:1011-1012

Brasaemle DL, Barber T, Wolins NE, Serrero G, Blanchette-Mackie EJ, Londos C (1997) Adipose differentiation-related protein is an ubiquitously expressed lipid storage droplet-associated protein. J Lipid Res 38:2249-2263.

Brown DA (2001) Lipid droplets: proteins floating on a pool of fat. Curr Biol 11:R446-449.

Chang BH, Li L, Paul A, Taniguchi S, Nannegari V, Heird WC, Chan L (2006) Protection against fatty liver but normal adipogenesis in mice lacking adipose differentiation-related protein. Mol Cell Biol 26:1063-1076

Chawla A, Lee C-H, Barak Y, He W, Rosenfeld J, Liao D, Han J, Kang H, and Evans RM (2003) PPARδ is a very low-density lipoprotein sensor in macrophages. Proc Natl Acad Sci USA 100:1268-1273.

Edvardsson U, Ljungberg A, Linden D, William-Olsson L, Peilot-Sjogren H, Ahnmark A, Oscarsson J () PPARα activation increases triglyceride mass and adipose differentiation-related protein in hepatocytes. J Lipid Res 47:329-340.

Fujimoto Y, Itabe H, Sakai J, Makita M, Noda J, Mori M, Higashi Y, Kojima S, Takano T (2004) Identification of major proteins in the lipid droplet-enriched fraction isolated from the human hepatocyte cell line HuH7. Biochim Biophys Acta 1644:47-59.

Larigauderie G, Furman C, Jaye M, Lasselin C, Copin C, Fruchart J-C, Castro G, Rouis M (2004) Adipophilin enhances lipid accumulation and prevents lipid efflux from THP-1 macrophages: potential role in atherogenesis. Arterioscler Thromb Vasc Biol 24:504-510.

Londos C, Brasaemle DL, Schultz CJ, Segrest JP, Kimmel AR. (1999) Perilipins, ADRP, and other proteins that associate with intracellular neutral lipid droplets in animal cells. Semin Cell Dev Biol 10:51-58.

Masuda Y, Itabe H, Odaki M, Hama K, Fujimoto Y, Mori M, Sasabe N, Aoki J, Arai H, Takano T (2006) ADRP/adipophilin is degraded through the proteasome-dependent pathway during regression of lipid-storing cells. J Lipid Res 47:87-98.

Targett-Adams P, McElwee MJ, Ehrenborg E, Gustafsson MC, Palmer CN, McLauchlan J (2005) A PPAR response element regulates transcription of the gene for human adipose differentiation-related protein. Biochim Biophys Acta. 1728:95-104.

Roles of Vasoactive Agents in Macrophage Foam Cell Formation and Atherosclerosis

Takuya Watanabe and Akira Miyazaki

Department of Biochemistry, Showa University School of Medicine
1-5-8 Hatanodai, Shinagawa-ku, Tokyo 142-8555, Japan.

Summary. Hypertension is a well-known risk factor for atherosclerosis, and vasoactive agents play a key role as a mediator between the two diseases. This study explored the role of human urotensin II (UII), the most potent vasoconstrictor to date, in atherosclerotic plaque formation in hypertensive patients, and the effects of UII and other vasoactive agents on foam cell formation and acyl-CoA:cholesterol acyltransferase-1 (ACAT1) expression in human monocyte-macrophages. Plasma UII levels were significantly higher in 50 patients with essential hypertension than in 31 normotensive controls (7.9 vs. 2.3 ng/ml), and correlated positively with systolic blood pressure, carotid artery intima-media thickness, and plaque score. In primary cultured human monocyte-macrophages, UII enhanced acetylated LDL-induced cholesteryl ester accumulation along with an increase in ACAT1 protein expression by ~2.5-fold. ACAT activity and ACAT1 mRNA levels were also increased. The ACAT1 expression increased by UII was completely abolished by UII receptor (UT) antagonists and inhibitors of G protein, c-Src tyrosine kinase, protein kinase C (PKC), extracellular signal-regulated kinase (ERK), and Rho kinase. UII stimulated ACAT1 expression at the highest level among vasoactive G-protein agonists such as endothelin-1, angiotensin II, serotonin, and salusin-β. These results suggest that UII accelerates foam cell formation by upregulating ACAT1 expression via UT/G protein/c-Src/PKC/ERK and Rho kinase pathways in human monocyte-macrophages, leading to the development of atherosclerotic plaque in essential hypertension.

Key words. vasoactive agents, acyl-CoA:cholesterol acyltransferase-1, macrophage foam cells, atherosclerosis, hypertension

1 Introduction

Hypertension is a well-known risk factor for atherosclerosis. Vasoconstrictor agents, such as urotensin II (UII), angiotensin II, endothelin-1, and serotonin, all of which are G-protein agonists, play a key role as a mediator linking hypertension and atherosclerosis (Watanabe et al., 2006a). UII is best known as the most potent vasoconstrictor to date (Ames et al., 1999). The contractile effect is great in order of potency as follows: UII, endothelin-1, angiotensin II or serotonin. Salusin-α and -β are newly identified 2 related polypeptides, and salusin-β induces the potent hypotension accompanied with bradycardia than does salusin-α (Shichiri et al., 2002). These vasoactive G-protein agonists are also mitogens for vascular smooth muscle cells (VSMCs) and fibroblasts. Therefore, these vasoactive agents are thought to participate in the progression of atherosclerotic plaques.

Recent studies have shown that the UII and its receptor (UT) system is involved in the pathogenesis of systemic and pulmonary hypertension. In blood vessels, UII is highly expressed in endothelial cells and lymphocytes whereas the high level of UT expression is observed in endothelial cells, VSMCs, monocytes, and macrophages (Watanabe et al., 2006b). The levels of UII and UT expression are upregulated by hypoxia and the inflammatory mediators such as interleukin-6 and 1β and interferon-γ. Plasma UII levels are increased in vascular endothelial dysfunction-related diseases, such as essential hypertension, ischemic heart disease, diabetes mellitus, heart failure, and renal failure (Watanabe et al., 2006b).

Accumulation of macrophage foam cells is a hallmark in the early stages of atherosclerotic lesions. Foam cells produce various bioactive molecules, such as cytokines, growth factors, and proteases, which play crucial roles in the progression of atherosclerosis. Intracellular free cholesterol is increased by the uptake of acetylated low-density lipoprotein (acetyl-LDL) via scavenger receptor class A (SR-A) and is decreased by the efflux of free cholesterol mediated by the ATP-binding cassette transporter A1 (ABCA1). Because excessive accumulation of free cholesterol is toxic to cells, free cholesterol must be either removed through efflux to extracellular acceptors such as high-density lipoprotein or esterified to cholesteryl ester (CE) by the microsomal enzyme acyl-CoA:cholesterol acyltransferase-1 (ACAT1). ACAT1 is expressed at high levels by macrophage foam cells in human atherosclerotic lesions *in vivo* (Miyazaki et al., 1998), and increases during differentiation from monocytes into macrophages *in vitro* (Suguro T et al., 2006). Therefore, ACAT1 plays a crucial role in the formation of macrophage foam cells. In primary cultured human monocyte-macrophages, ACAT1 expression is upregulated by dexamethasone, dehydroepiandrosterone, and transforming growth factor-β_1, but is down-regulated by adiponectin (Watanabe et al., 2006b). However, it has not been clarified whether vasoactive agents modulate ACAT1 expression followed by foam cell formation in human monocyte-macrophages.

In the present study, we assessed the effects of UII on the expression of ACAT1 and ABCA1, activities of ACAT and SR-A, and acetyl-LDL-induced CE accumulation in primary human monocyte-macrophages. Further, we compared the ACAT1 expression enhanced by UII with that by other vasoactive G-protein agonits, such as angiotensin II, endothelin-1, serotonin, salusin-α, and salusin-β.

2 UII Increased ACAT1 Expression and Acetyl-LDL-Induced CE Accumulation in Human Monocyte-Macrophages

Seven days after primary monocytic culture, UII increased ACAT1 protein expression in a concentration-dependent manner, with the maximal increase of 2.5-fold observed at 25 nM (Fig. 1A). The ACAT1 protein expression increased by UII was completely abolished by selective UT antagonists or specific inhibitors of G protein, c-Src tyrosine kinase, protein kinase C (PKC), extracellular signal-regulated kinase (ERK), and Rho kinase (ROCK) (Watanabe et al., 2005). UII also increased ACAT enzyme activity in parallel with ACAT1 protein expression (Fig. 1B). Northern blotting analysis revealed that UII (25 nM) significantly increased ACAT1 mRNA levels of the 2.8- and 3.6-kb transcripts by ~1.8-fold, but had no significant effect on 4.3- or 7.0-kb transcripts (Fig. 1C). In addition to ACAT1 expression, UII (25 nM) significantly accelerated acetyl-LDL-induced CE accumulation (Fig. 1D). However, UII had no significant effects on SR-A function as assessed by endocytic uptake of [^{125}I]acetyl-LDL (Fig. 1E) and ABCA1 protein expression (Fig. 1F).

3 Effects of Serotonin and Other Vasoactive Agents on ACAT1 Expression in Human Monocyte-Macrophages

Serotonin (10 μM) increased ACAT1 protein expression by 2-fold, which was completely inhibited by the pretreatment with 5-HT$_{2A}$ antagonist sarpogrelate and its metabolite M-1 [Suguro et al., 2006]. ACAT1 protein expression was significantly increased by angiotensin II (2-fold) but not by endothelin-1 (1.4-fold) [Watanabe et al. 2006b]. We also tested the effect of salusin-α and salusin-β, which are the different parts of the same precursor peptide, so-called prosalusin. Salusin-β increased ACAT1 protein expression by 2.2-fold at 0.6 nM in a maximum, whereas salusin-α decreased it in a concentration-dependent manner. Collectively, the ACAT1 protein expression by UII is the highest level among those by other vasoactive G-protein agonists, such as serotonin, angiotensin II, endothelin-1, salusin-α and -β.

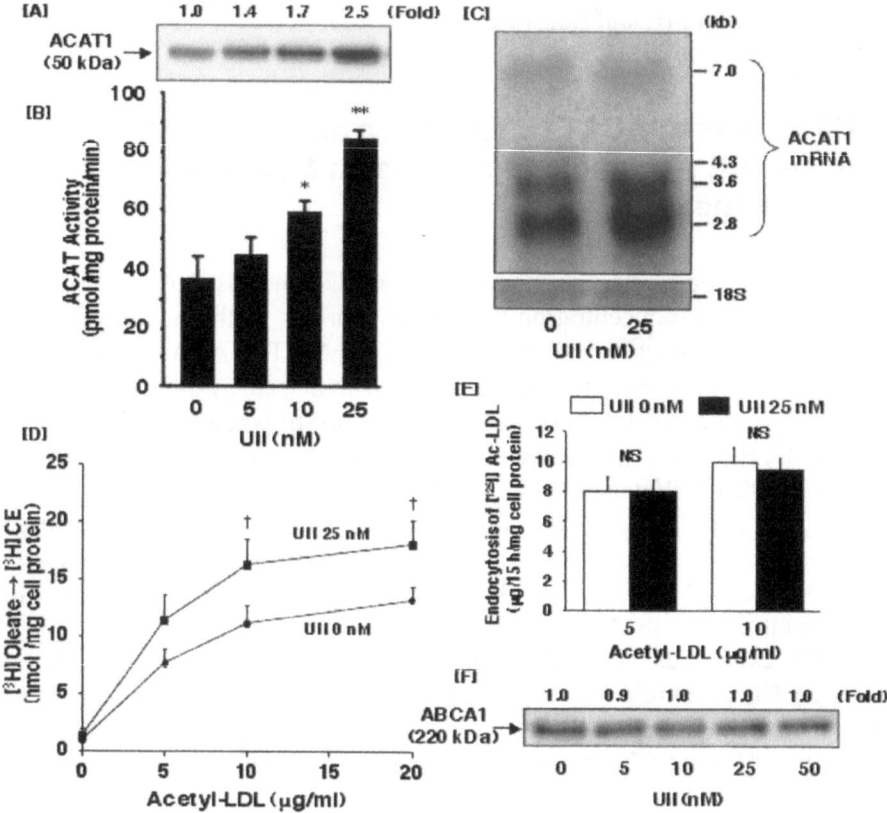

Fig. 1. Effects of urotensin II (UII) on acyl-CoA:cholesterol acyltransferase-1 (ACAT1) expression, ACAT activity, acetylated low-density lipoprotein (acetyl-LDL)-induced cholesteryl ester (CE) accumulation, endocytic uptake of [^{125}I]acetyl-LDL, and ATP-binding cassette transporter A1 (ABCA1) expression in human monocyte-macrophages. Human monocytes were incubated for 7 days with the indicated concentrations of UII. ACAT1 [A] and ABCA1 [F] protein, ACAT mRNA [C] expression, and ACAT enzyme activity [B] were determined by Western or Northern blotting analyses and the reconstituted assay, respectively. Otherwise, the remaining cells were further incubated for 24 h with the indicated concentrations of acetyl-LDL in the presence of 0.1 mM [^{14}C]oleate followed by determination of cellular CE accumulation by the radioactivity of cholesteryl [^{14}C]oleate [D], or for 15 h with the indicated concentrations of [^{125}I]acetyl-LDL followed by determination of degradation of [^{125}I]acetyl-LDL [E]. Data are expressed as the mean ± SE from 3 independent experiments using monocytes from 3 different donors. *$p < 0.0005$, †$p < 0.05$ vs. UII 0 nM. **$p < 0.0001$ vs. other concentrations of UII.

4 Relation Between Plasma UII Levels and Carotid Atherosclerosis in Hypertensive Patients

We studied the correlation of plasma UII levels with systolic blood pressure (BP) and atherosclerosis as evaluated by intima-media thickness (IMT) and plaque score (the sum of all plaque thicknesses) in the carotid artery on ultrasonography. Plasma UII levels were significantly higher in 50 patients with essential hypertension than in 31 normotensive controls (7.9 ± 1.1 vs. 2.3 ± 0.2 ng/ml, $p < 0.0001$), and correlated positively with systolic BP, the maximum IMT, and plaque score (Fig. 2A–C). Plasma U-II levels increased significantly in accordance with the severity of grade of plaque score (Fig. 2D) [Suguro et al., 2007]. In addition, the expression of UII was observed at high levels in atherosclerotic plaques from the carotid artery of hypertensive patient.

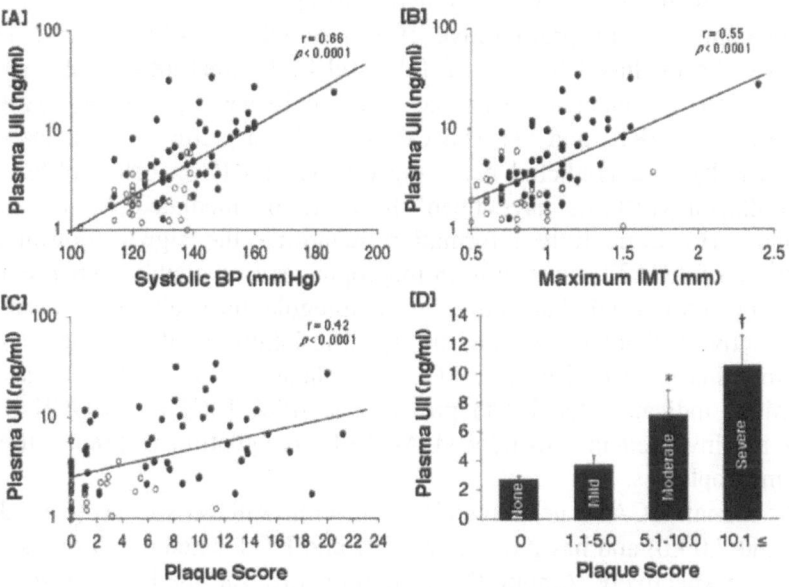

Fig. 2. Correlations of plasma urotensin II (UII) levels with systolic blood pressure (BP) [A], maximum intima-media thickness (IMT) [B], and plaque score [C] in hypertensive patients (n = 50, closed circles) and normotensive controls (n = 31, open circles). [D] Plasma UII levels increased significantly in accordance with the severity of grade of plaque score (the sum of all plaque thicknesses) in all the subjects (n = 81). Data are shown as the mean ± SE. *$P < 0.05$, †$P < 0.0001$ vs. plaque score 0 (None).

5 Discussion

The present study shows that, among vasoactive G-protein agonists, UII, serotonin, angiotensin II, and salusin-β, but not endothelin-1, could upregulate significantly ACAT1 expression in primary human monocyte-macrophages. We regard 2-fold increase in ACAT1 expression by these vasoactive agents as a significant cellular event that accelerates the formation of human macrophage foam cells (Watanabe et al., 2005). UII stimulated ACAT1 expression at the highest level among these vasoactive G-protein agonists. In contrast, salusin-α significantly reduced ACAT1 expression in human monocyte-macrophages. In addition, the increase in acetyl-LDL-induced CE accumulation by UII (~1.4-fold) can be regarded as a significant change because it is comparable to the previous reports that showed ~1.2-fold increase by dehydroepiandrosterone (Ng et al., 2003) and ~1.6-fold increase by high glucose (Fukuhara-Takaki et al., 2005). Salusin-β increased acetyl-LDL-induced CE accumulation by ~1.5-fold, whereas salusin-α decreased it in a half (data not shown).

The contractile or proliferative effects of UII on VSMCs have been shown to be mediated by the UT followed by various intracellular signal transduction mechanisms, such as phospholipase C, protein tyrosine kinases, PKC, ERK, and the RhoA/ROCK-related pathways (Watanabe et al., 2001a,b). In endothelial cells, the effects of UII on cell proliferation and collagen synthesis have been shown to be mediated via the ERK pathway. However, little information regarding the signal transduction pathways of ACAT1 expression in macrophages is available. Our previous study showed that UII-induced ACAT1 upregulation is abolished by using the selective UT antagonists and the specific inhibitors of G protein, c-Src tyrosine kinase, PKC, ERK, or ROCK [Watanabe et al., 2005]. These observations indicated that UT/G protein/c-Src/PKC/ERK and ROCK pathways are involved in UII-induced ACAT1 upregulation in human monocyte-macrophages.

The human ACAT1 gene encodes 4 different mRNA species (2.8-, 3.6-, 4.2-, and 7.0 kb) and has 2 promoters (P1 and P7) located on chromosomes 1 and 7, respectively. Among the 4 ACAT1 mRNAs, the levels of the 2.8- and 3.6-kb transcripts were selectively increased by UII, whereas the 4.3- and 7.0-kb transcripts remained unchanged. The 2.8- and 3.6-kb transcripts are regulated by the P1 promoter, while the 4.3-kb transcript is regulated by both P1 and P7 promoters. The present study demonstrated increases in expression of 2 shorter transcripts by UII during differentiation of monocytes into macrophages. The same phenomenon was observed when ACAT1 was upregulated by transforming growth factor-β1, serotonin, and angiotensin II. Our experiments demonstrated that the ACAT1 upregulation by UII occured predominantly during differentiation of human monocytes into macrophages rather than after differentiation (Watanabe et al., 2005).

The present study showed that plasma UII levels were significantly increased in patients with essential hypertension as compared with normotensive controls. Furthermore, plasma UII levels showed significant positive correlations with systolic BP, carotid IMT, plaque score, and homeostasis model assessment for insulin resistance (HOMA-IR). Multiple logistic regression analysis revealed that plasma level of UII was significantly associated with a 1.6-fold higher risk of carotid plaque formation (plaque score $\geq$ 1.1) as compared with other widely accepted risk factors, such as age, systolic BP, serum levels of small dense LDL and high-sensitive C-reactive protein, and HOMA-IR (Suguro et al., 2007).

6 Conclusions

The results of the present study suggest that UII plays a novel role in macrophage foam cell formation by upregulating ACAT1 expression via the UT/G protein/c-Src/PKC/ERK and ROCK pathways, but not by SR-A or ABCA1. UII stimulated ACAT1 expression at the highest level among vasoactive agents, such as endothelin-1, angiotensin II, serotonin, and salusin-β. These findings provide an understanding of potential molecular mechanisms by which hypertension promotes the development of atherosclerosis. Further, our results suggest that the UII/UT system may be a promising new therapeutic target in atherosclerotic vascular diseases.

7 References

Ames RS, Sarau HM, Cambers JK, Willette RN, Aiyar NV, Romanic AM, Louden CS, Foley JJ, Sauermelch CF, Coatney RW, Ao Z, Disa J, Holmes SD, Stadel JM, Martin JD, Liu WS, Glover GI, Wilson S, McNulty DE, Ellis CE, Elshourbagy NA, Shabon U, Trill JJ, Hay DWP, Ohlstein EH, Bergsma DJ, Douglas SA (1999) Human urotensin II is a potent vasoconstrictor and agonist for the orphan receptor GPR14. Nature 401:282-286

Fukuhara-Takaki K, Sakai M, Sakamoto Y, Takeya M, Horiuchi S (2005) Expression of class A scavenger receptor is enhanced by high glucose in vitro and under diabetic conditions in vivo. J Biol Chem 280:3355-3364

Ng MKC, Nakhla S, Baoutina A, Jessup W, Handelsman DJ, Celermajer DS (2003) Dehydroepiandrosterone, an adrenal androgen, increases human foam cell formation. J Am Coll Cardiol 42:1967-1974

Miyazaki A, Sakashita N, Lee O, Takahashi K, Horiuchi S, Hakamata H, Morganelli PM, Chang CC, Chang TY (1998) Expression of ACAT-1 protein in human atherosclerotic lesions and cultured human monocytes-macrophages. Arterioscler Thromb Vasc Biol 18:1568-1574

Shichiri M, Ishimaru S, Ota T, Nishikawa T, Isogai T, Hirata Y (2003) Salusins: newly identified bioactive peptides with hemodynamic and mitogenic activities. Nat Med 9:1166-1172

Suguro T, Watanabe T, Kanome T, Kodate S, Hirano T, Miyazaki A, Adachi M (2006) Serotonin acts as an up-regulator of acyl-coenzyme A:cholesterol acyltransferase-1 in human monocyte-macrophages. Atherosclerosis 186:275-281

Suguro T, Watanabe T, Ban Y, Kodate S, Misaki A, Hirano T, Miyazaki A, Adachi M (2007) Increased human urotensin II levels are correlated with carotid atherosclerosis in essential hypertension. Am J Hypertens 20:211-217

Watanabe T, Pakala R, Katagiri T, Benedict CR (2001a) Synergistic effect of urotensin II with mildly oxidized LDL on DNA synthesis in vascular smooth muscle cells. Circulation 104:16-18

Watanabe T, Pakala R, Katagiri T, Benedict CR (2001b) Synergistic effect of urotensin II with serotonin on vascular smooth muscle cell proliferation. J Hypertens 19:2191-2196

Watanabe T, Suguro T, Kanome T, Sakamoto Y, Kodate S, Hagiwara T, Hongo S, Hirano T, Adachi M, Miyazaki A (2005) Human urotensin II accelerates foam cell formation in human monocyte-derived macrophages. Hypertension 46:738-744

Watanabe T, Kanome T, Miyazaki A (2006a) Relationship between hypertension and atherosclerosis: from a viewpoint of the most potent vasoconstrictor human urotensin II. Curr Hypertens Rev 2:237-246

Watanabe T, Kanome T, Miyazaki A, Katagiri T (2006b) Human urotensin II linking hypertension and coronary artery disease. Hypertens Res 29:375-387

Part III
Risk Factors for Cardiovascular Diseases

Sleep Apnea and Metabolic and Cardiovascular Complications

Christopher P. O'Donnell

Division of Allergy, Pulmonary, and Critical Care Medicine, Department of Medicine, University of Pittsburgh, NW628 MUH 3459 Fifth Ave, Pittsburgh, PA 15213, USA

Summary. Obesity is reaching epidemic proportions in adults and its prevalence is rising dramatically in children. Obesity, especially central obesity, leads to insulin resistance and type 2 diabetes mellitus, affecting more than 100 million people worldwide. Central obesity is also a major risk factor for sleep apnea, which affects 2-4% of adults in the United States, and has prevalence in excess of 50% in obese, otherwise healthy, males. Thus, both metabolic dysfunction and sleep apnea (SA) are adverse outcomes of obesity and act as important intermediates in the path to cardiovascular morbidity and mortality. The fact that SA and diabetes frequently co-exist in obese individuals has received growing attention. The overwhelming majority of large community and clinic-based studies report a positive association between SA and one or more parameters of metabolic dysfunction, independent of obesity. Treatment of SA with nasal continuous positive airway pressure can lead to improvements in insulin sensitivity, in the absence of any change in body weight, and animal models of SA exhibit reduced insulin sensitivity suggesting that a cause and effect relationship exists between sleep apnea and insulin resistance. Sleep apnea is characterized by recurrent collapse of the upper airway during sleep leading to periods of intermittent hypoxia and sleep fragmentation. The stimulus of intermittent hypoxia is known to activate multiple physiologic systems including sympathetic nerve activity (SNA), the hypothalamic-pituitary-adrenal axis, insulin counter-regulatory hormones, a stress/pro-inflammatory state, and generation of reactive oxygen species. All of these activated pathways can contribute to the development of insulin resistance and putatively act as intermediates in the pathway leading from sleep apnea to cardiovascular morbidity and mortality.

Key words. corticosterone, insulin resistance, intermittent hypoxia, sympathetic nerve activity

1 Sleep Apnea, Obesity, and Cardiovascular Morbidity and Mortality

In the last two decades an epidemic in obesity has occurred in the United States and more than half the adult population is now considered overweight or obese (National Institutes of Health, 1998). The dramatic increase in the prevalence of obesity represents an enormous public health burden now and in the future, as obesity is insidiously permeating into the childhood years. Obesity is also associated with broad ranging cardiovascular pathophysiology such as hypertension, ventricular hypertrophy, coronary artery disease, and stroke. In addition, there is increasing recognition that obesity can cause respiratory problems, in particular SA (Young et al. 1993), which itself has been shown to cause serious cardiovascular pathophysiology. Consequently, SA represents an important link between obesity and cardiovascular disease.

The more than 50% prevalence of SA in obese individuals (Punjabi et al. 2002) indicates that SA contributes significantly to the adverse cardiovascular outcomes associated with obesity. The fact that SA can elevate SNA (Narkiewicz et al. 1998), and cause daytime hypertension and cardiac hypertrophy independent of body weight (Peppard et al. 2000, Guidry et al. 2001), suggests that the interaction of obesity and SA represents a serious cardiovascular risk. A recent prospective epidemiologic study (Mooe et al. 2001), clearly demonstrates that the presence of SA in patients with coronary artery disease significantly impacts on mortality. Other studies have shown that SA increases the risk for sudden cardiac death at nighttime (Gami et al. 2005). A long-term prospective study has also shown that the cumulative incidence of fatal and non-fatal cardiovascular events is significantly elevated in the patients with severe SA (Marin et al. 2005). Moreover the incidence of fatal and non-fatal cardiovascular events is reduced in patients undergoing treatment for their SA with nasal continuous positive airway pressure [CPAP] (Marin et al. 2005). Finally, SA was the predominant risk factor for stroke or death from any cause in a recent study that included assessment of hypertension, hyperlipidemia, artrial fibrillation, body mass index and smoking history (Yaggi et al. 2005). Thus,

SA increases the risk of hypertension, most likely through an elevation in SNA, and also increases the risk of stroke and sudden death at night, and the incidence of non-fatal and fatal cardiac events.

2 Sleep Apnea, Obesity and Insulin Resistance

Metabolic dysfunction is potentially an intermediate pathway contributing to the development of cardiovascular morbidity and mortality. The fact that SA and diabetes frequently co-exist in obese individuals has received growing attention in the clinical literature (Punjabi et al. 2002). Some investigators attributed this coexistence to obesity being a common risk factor for both SA and diabetes, whereas others have suggested that there are physiological links between SA and insulin resistance, independent of obesity. We previously undertook a systematic review of the literature on the relationship between SA and glucose intolerance and insulin resistance (Punjabi et al. 2003). The overwhelming majority of large community and clinic-based studies reported a positive association between SA and one or more parameters of metabolic dysfunction, independent of obesity. Moreover, there was evidence of an association between glucose intolerance and insulin resistance and the presence of hypoxic stress in SA patients. In contrast, several small interventional studies were unable to demonstrate any significant metabolic improvements with the application of continuous positive airway pressure (CPAP) to abolish SA. However, a recent study (Harsch et al. 2004) conducted serial measurements of insulin sensitivity using the gold standard of the hyperinsulinemic euglycemic clamp in forty SA patients before and after treatment with CPAP. They observed significant improvements in insulin sensitivity with CPAP treatment in apneic patients. Furthermore, they made the interesting observation that the less obese patients had the greatest improvements in insulin sensitivity with CPAP treatment. However, to this point there has been no direct cause and effect evidence linking SA to insulin resistance.

We have recently undertaken a series of studies to examine the effects of intermittent hypoxia (IH), the predominant physiological disturbance in SA, on the development of insulin resistance in a mouse model (Iiyori et al. 2007). We utilized the most sensitive measure of insulin sensitivity, the hyperinsulinemic euglycemic clamp, and exposed mice to 9 hours of IH (60 events per hour, nadir $FIO2 = 5\text{-}6\%$). Whole body insulin sensitivity was reduced by 21.5% (49.4 ± 1.5 ml/kg/min to 38.8 ± 2.7 ml/kg/min) during exposure to IH in lean, otherwise healthy, C57BL/6J mice. Moreover, when glucose uptake was specifically determined in muscle, only the predominantly oxidative, insulin-sensitive soleus muscle showed marked decreases in glucose utilization in response to IH exposure. In contrast, baseline hepatic glucose output was unaffected by IH. Thus, in the absence of obesity, acute exposure to IH can significantly reduce insulin sensitivity, specifically in oxidative muscle fibers.

3 Putative Mechanisms for Sleep Apnea to Cause Insulin Resistance

The two most immediate physiological consequences of upper airway collapse in SA are rapid arterial oxyhemoglobin desaturation and arousal from sleep. The need to maintain appropriate blood gas levels and sleep represent basic homeostatic drives, and it is not surprising that deficits in either can produce diverse physiological and biochemical consequences. Several studies have demonstrated that SA causes acute increases in SNA (Leuenberger et al. 1995, Somers et al. 1995). The SNA increases acutely with each apneic episode and the effect carries over from the nighttime to produce elevated daytime levels of SNA in SA patients (Leuenberger et al. 1995, Somers et al. 1995). It is likely that both the hypoxic and arousal components of SA elevate SNA. We have shown in clinical studies that acute increases in peripheral vascular tone during episodes of SA are dependent on the degree of arterial hemoglobin desaturation and the presence of an identifiable arousal response (O'Donnell et al. 2002). There is now a consensus in the literature that human SA and animal models of IH exhibit increased levels of SNA and can cause sustained hypertension.

Activation of the sympathoadrenal axis by either IH or SF may also impact negatively on metabolic function in susceptible individuals with obesity and SA. In the short-term, activation of the sympathoadrenal axis will preferentially increase glucose uptake to the brain and heart by reducing insulin sensitivity in peripheral tissues, inhibiting insulin secretion from the pancreas, and increasing glucose output from the liver. Given that SNA and cortisol levels are chronically elevated in SA during both the nighttime apneic episodes and the subsequent daytime period (Leuenberger et al. 1995, Somers et al. 1995), the sympathoadrenal axis could significantly contribute to the increased insulin resistance reported in this patient population.

On the other hand, recent evidence suggests that activation of some neural pathways may also improve insulin sensitivity. For example, administration of α-melanocyte-stimulating hormone causes an increase in insulin sensitivity in peripheral tissues. The peripheral action of central α-MSH is dependent on sympathetic activation of β_3-adrenergic receptors on adipocytes. The stimulation of β_3-adrenergic receptors reduces circulating free fatty acid levels and increases activity of uncoupling protein 1 in adipocyctes, both of which will increase insulin-mediated uptake of glucose. Thus, an increase in SNA as a result of SA may increase or decrease insulin sensitivity depending on the subtype of adrenergic receptor that is stimulated. No clinical studies have yet determined the impact of the sympathoadrenal axis on insulin sensitivity in SA.

However, in our recent study of IH in mice we examined the effect of blocking the autonomic nervous system on changes in insulin sensitivity (Iiyori et al. 2007). The insulin sensitivity was assessed by the hyperinsulinemic euglycemic clamp during IH with a constant infusion of hexamethonium (10 mg/kg bolus + 20 mg/kg/hr infusion) to produce blockade of the autonomic ganglia by occupying cholinergic receptor sites on the postsynaptic membrane. Despite hexamethonium causing hypotension and a lowering of baseline blood glucose, insulin sensitivity determined by the clamp was still significantly reduced during exposure to IH. These data suggest that at least under conditions of shorterm exposure to IH, activation of the sympathoadrenal axis is not required to elicit reductions in insulin sensitivity.

Potentially SA or IH may decrease insulin sensitivity by increasing levels of other counter-regulatory hormones. Among other counter-regulatory hormones, there is data from one clinical study indicating that cortisol is elevated in SA (Bratel et al. 1999). Moreover, a recent study in normal individuals demonstrated that six consecutive nights of sleep restriction caused a decrease in glucose tolerance that was associated with elevated cortisol levels (Spiegel et al. 1999). If the sleep debt incurred with sleep restriction in any way mimics the sleep disruption of SA, then elevated cortisol levels in SA patients may contribute to glucose intolerance and insulin resistance in this patient population. Data from our studies of IH in mice show that exposure to hypoxia causes an increase in corticosterone (Iiyori et al. 2007), the predominant glucocorticoids in rodents. Thus, increased secretion of cortisol may at least in part contribute to reducing insulin sensitivity in SA. The impact of other counter-regulatory hormones, such as growth hormone and glucagons, on insulin sensitivity in SA or IH has not as yet been studied.

The recent surge in obesity has resulted in a shift of emphasis away from counter-regulatory hormones to a focus on lipotoxicity as a mediator of insulin resistance. Liptoxicity can result in hyperlipidemia, ectopic fat deposits and increased inflammatory cytokines, leading to generation of reactive oxygen species and activitation of pro-inflammatory/stress pathways, which can cause insulin resistance. Interestingly, many of these pathways of lipotoxicity are known to be activated in response to SA or IH in the absence of comorbidites associated with obesity (Yokoe et al, 2003, Li et al. 2005, Minoguchi et al. 2006). At present we know little about whether lipotoxicity pathways can induce insulin resistance in patients with SA or in rodent models of IH. However, the potential for two independent factors, obesity and SA, to activate lipotoxic pathways may have important clinical consequences and contribute to the established relationship between SA and cardiovascular morbidity and mortality.

References

Bratel T, Wennlund A, Carlstrom K (1999) Pituitary reactivity, androgens and catecholamines in obstructive sleep apnoea. Effects of continuous positive airway pressure treatment (CPAP). Respir Med 93:1-7

Gami AS, Howard DE, Olson EJ, Somers VK (2005) Day-night pattern of sudden death in obstructive sleep apnea. N Engl J Med 352:1206-1214

Guidry UC, Mendes LA, Evans JC, Levy D, O'Connor GT, Larson MG, Gottlieb DJ, Benjamin EJ (2001) Echocardiographic features of the right heart in sleep-disordered breathing: the Framingham Heart Study. Am J Respir Crit Care Med 164:933-938

Harsch IA, Schahin SP, Radespiel-Troger M, Weintz O, Jahreiss H, Fuchs FS, Wiest GH, Hahn EG, Lohmann T, Konturek PC, Ficker JH (2004) Continuous positive airway pressure treatment rapidly improves insulin sensitivity in patients with obstructive sleep apnea syndrome. Am J Respir Crit Care Med 169:156-162

Iiyori N, Alonso LC, Li J, Sanders MH, Garcia-Ocana A, O'Doherty RM, Polotsky VY, O'Donnell CP (2007) Intermittent hypoxia causes insulin resistance in lean mice independent of autonomic activity. Am J Respir Crit Care Med 175:851-857

Leuenberger U, Jacob E, Sweer L, Waravdekar N, Zwillich C, Sinoway L (1995) Surges of muscle sympathetic nerve activity during obstructive apnea are linked to hypoxemia. J Appl Physiol 79:581-588

Li J, Thorne LN, Punjabi NM, Sun CK, Schwartz AR, Smith PL, Marino RL, Rodriguez A, Hubbard WC, O'Donnell CP, Polotsky VY (2005) Intermittent hypoxia induces hyperlipidemia in lean mice. Circ Res 97:698-706

Marin JM, Carrizo SJ, Vicente E, Agusti AG (2005) Long-term cardiovascular outcomes in men with obstructive sleep apnoea-hypopnoea with or without treatment with continuous positive airway pressure: an observational study. Lancet 365:1046-1053

Minoguchi K, Yokoe T, Tanaka A, Ohta S, Hirano T, Yoshino G, O'Donnell CP, Adachi M (2006) Association between lipid peroxidation and inflammation in obstructive sleep apnoea. Eur Respir J 28:378-385

Mooe T, Franklin KA, Holmstrom K, Rabben T, Wiklund U (2001) Sleep-disordered breathing and coronary artery disease: long-term prognosis. Am J Respir Crit Care Med 164:1910-1913

Narkiewicz K, van de Borne PJ, Cooley RL, Dyken ME, Somers VK (1998) Sympathetic activity in obese subjects with and without obstructive sleep apnea. Circulation 98:772-776

National Institutes of Health (1998) Clinical guidelines on the identification, evaluation, and treatment of overweight and obesity in adults-the evidence report. Obes Res 6 (Suppl 2):51S-209S

O'Donnell CP, Allan L, Atkinson P, Schwartz AR (2002) The effect of upper airway obstruction and arousal on peripheral arterial tonometry in obstructive sleep apnea. Am J Respir Crit Care Med 166:965-971

Peppard PE, Young T, Palta M, Skatrud J (2000) Prospective study of the association between sleep-disordered breathing and hypertension. N Engl J Med 342:1378-1384

Punjabi NM, Ahmed MM, Polotsky VY, Beamer BA, O'Donnell CP (2003) Sleep-disordered breathing, glucose intolerance, and insulin resistance. Respir Physiol Neurobiol 136:167-178

Punjabi NM, Sorkin JD, Katzel LI, Goldberg AP, Schwartz AR, Smith PL (2002) Sleep-disordered breathing and insulin resistance in middle-aged and overweight men. Am J Respir Crit Care Med 165:677-682

Somers VK, Dyken ME, Clary MP, Abboud FM (1995) Sympathetic neural mechanisms in obstructive sleep apnea. J Clin Invest 96:1897-1904

Spiegel K, Leproult R, Van Cauter E (1999) Impact of sleep debt on metabolic and endocrine function. Lancet 354:1435-1439

Yaggi HK, Concato J, Kernan WN, Lichtman JH, Brass LM, Mohsenin V (2005) Obstructive sleep apnea as a risk factor for stroke and death. N Engl J Med 353:2034-2041

Yokoe T, Minoguchi K, Matsuo H, Oda N, Minoguchi H, Yoshino G, Hirano T, Adachi M (2003) Elevated levels of C-reactive protein and interleukin-6 in patients with obstructive sleep apnea syndrome are decreased by nasal continuous positive airway pressure. Circulation 107:1129-1134

Young T, Palta M, Dempsey J, Skatrud J, Weber S, Badr S (1993) The occurrence of sleep-disordered breathing among middle-aged adults. N Engl J Med 328:1230-1235

Sleep Apnea Syndrome and Atherosclerosis

Kenji Minoguchi

First Department of Internal Medicine, Showa University School of Medicine, 1-5-8 Hatanodai, Shinagawa-ku, Tokyo, 142-8666, Japan

Summary. Obstructive sleep apnea (OSA) is a risk factor for a variety of cardiovascular diseases. Recent studies suggest that OSA is also independently associated with atherosclerosis. The intermittent hypoxia occurring in OSA patients has been proposed to represent a form of oxidative stress that causes generation of ROS and the development of systemic inflammation. We have reported that urinary levels of 8-isoprostane, a marker for oxidative stress, and serum levels of high-sensitivity C-reactive protein, interleukin-6, tumor necrosis factor-α, and matrix metalloproteinsase-9, all markers for atherosclerosis, are elevated in patients with OSA. Moreover, carotid intima-media thickness and the percentage of silent brain infarction were significantly increased in patients with OSA compared to obese control subjects, consistent with the presence of atherosclerosis. Therefore, these results suggest that increased generation of ROS and development of systemic inflammation likely contribute to the progression of atherosclerosis in OSA.

Key words. Sleep apnea, inflammation, atherosclerosis, cytokine, hypoxia

1 OSA and Systemic Inflammation

Ongoing inflammatory responses play important roles in atherosclerosis (Glass et al. 2001). Although C-reactive protein (CRP) is a nonspecific marker of inflammation, recent epidemiological studies suggest that CRP is an important risk factor in atherosclerosis and coronary artery diseases

(Haverkate et al. 1997; Lindahl et al. 2000). CRP directly induces adhesion molecules on endothelial cells and chemokine production by human umbilical vein endothelial cells (Pasceri et al. 2000; Pasceri et al. 2001). Interleukin (IL)-6 is a proinflammatory cytokine that is also implicated in the pathogenesis of atherosclerosis (Yudkin et al. 2000). Plasma levels of IL-6 are reportedly correlated with the mortality rate in patients with unstable coronary artery disease and with the risk of future myocardial infarction in apparently healthy men (Ridker PM et al. 2000). Therefore, serum levels of CRP and IL-6 and IL-6 production from monocytes were investigated at 5 a.m. in patients with OSA (Yokoe et al. 2003). Serum levels of CRP and IL-6 were significantly higher in patients with OSA than in obese control subjects. IL-6 production by monocytes was also higher in patients with OSA than in obese control subjects. Nasal continuous positive pressure (nCPAP) significantly decreased levels of CRP and IL-6 and production of IL-6 by monocytes. Therefore, OSA is associated with systemic inflammation with increased risks of cardiovascular morbidity and mortality and nCPAP may be useful for decreasing these risks.

2 OSA and Tumor Necrosis Factor (TNF)-α

TNF-α is a pleiotropic proinflammatory cytokine that exerts multiple biological effects. Early atherosclerosis is characterized by monocytes infiltrating into the vascular wall and later transforming foam cells. Several studies indicated that TNF-α is involved in each step of atherosclerosis by inducing such adhesion molecules as intracellular adhesion molecule -1 and vascular cell adhesion molecule-1, stimulating production of monocyte chemoattractant protein-1 by endothelial cells, promoting the proliferation and migration of smooth muscle cells, and inducing the expression of lectin-like oxidized low-density lipoprotein receptor-1 (Glass et al. 2001; Libby et al. 1999; Barks et al. 1997; Hatakeyama et al. 2002; Kume et al. 1998). In addition, plasma levels of TNF-α are associated with atherosclerosis. Therefore, we have investigated the serum levels of TNF-α and production of TNF-α by monocytes in patients with OSA (Minoguchi et al. 2004). Serum levels of TNF-α were significantly higher in patients with OSA than in obese control subjects. TNF-α production by monocytes was also higher in patients with OSA than in obese control subjects. nCPAP significantly decreased levels of TNF-α and production of TNF-α by monocytes. Therefore, TNF-α is elevated in patients with OSA and nCPAP may be useful for decreasing these levels.

3 OSA and Oxidative Stress

OSA is characterized by repetitive periods of upper airway collapse and results in cyclic periods of hypoxia/reoxygeneration that cause the increased generation of oxygen species by oxidative stress (Dean et al. 1993; Lavie et al. 2003). In fact, production of oxygen species from neutrophils and monocytes obtained from patients with OSA was increased and treatment with nCPAP significantly decreased the production of oxygen species from these cells (Schultz et al. 2000; Dyugovskaya et al. 2002). Moreover, oxidative damage is involved in the pathogenesis of atherosclerosis and cardiovascular diseases (Harrison et al. 2003; Dhalla et al. 2000). Therefore, oxidative stress may contribute to the cardiovascular risk profile in patients with OSA. In the cardiovascular system, lipids are in the first line of radical attack. 8-isoprostane, 8-iso-prostaglandin F2α, is one of the lipid markers for oxidative stress (Lawson et al. 1999). Due to its chemical stability, urinary excretion of 8-isoprostane is considered a reliable index of oxidant stress and ensuing lipid peroxidation in vivo. We have found that urinary excretion of 8-isoprostane was significantly higher in patients with moderate to severe OSA than in patients with mild OSA, obese subjects, or healthy subjects (Minoguchi et al. 2006). The severity of OSA was an independent factor predicting the urinary excretion of 8-isoprostane. nCPAP significantly decreased urinary excretion of 8-isoprostane and serum levels of hsCRP. Therefore, these results suggest that oxidative stress is increased in patients with OSA.

4 OSA and Atherosclerosis

Carotid intima-media thickness (IMT) is a useful marker for investigating the degree of early atherosclerosis (O'Leary et al. 2002). Several clinical studies have demonstrated that an increase in ultrasonographically measured carotid IMT is associated with elevated risks of cardiovascular diseases and stroke (O'Leary et al. 1999; Hodis et al. 1998; Cao et al. 2003; Magyar et al. 2003). In addition, carotid IMT is associated with such inflammatory markers for atherosclerosis as CRP, IL-6, and IL-18 (Wang et al. 2002; Leonsson et al. 2003; Aso et al. 2003). Moreover, carotid IMT is higher in patients with OSA than in obese control subjects (Silvestrini et al. 2002; Suzuki et al. 2004; Kaynak et al. 2003; Shultz et al. 2005). Therefore, OSA is independently associated with the progression of athersosclerosis. However, the relationship between carotid IMT and levels of inflammatory markers for atherosclerosis have not been previously studied in patients with OSA. We have studied the subjects who were free from other diseases and were taking no medications were studied to avoid the influence of confounding diseases. Carotid IMT of patients with OSA were significantly higher than those of obese control subjects (Minoguchi et al. 2005). Carotid IMT was significantly correlated with serum levels of CRP, IL-6, and IL-18, duration of OSA-related hypoxia, and severity of OSA. In addition, the primary factor influencing carotid IMT was duration of hypoxia during total sleep time. These results suggest that OSA-related hypoxia and systemic inflammation might be associated with the progression of atherosclerosis in patients with OSA.

Obstructive sleep apnea (OSA) is associated with increased cerebrovascular morbidity and mortality (Arzt et al. 2005; Yaggi et al. 2005). The occurrence of stroke in OSA patients is likely preceded by subclinical cerebrovascular disease, often termed silent brain infarction (SBI) that is detectable with brain magnetic resonance imaging. However, the independent effects of OSA on the prevalence of SBI have not been clearly established. In our study the percentage of SBI in patients with moderate to severe OSA (25.0%) was higher than that of obese control subjects (6.7%) or patients with mild OSA (7.7%) (Minoguchi et al. 2007). These results suggest that SBI is more common in patients with moderate to severe OSA, leading to elevated cerebrovascular morbidity.

5 OSA and Metalloproteinsase (MMP)-9

Matrix metalloproteinases (MMPs) are a family of zinc-containing endoproteases that share structural domains but differ in substrate specificity, cellular sources, and inducibility (Visse et al. 2003). The expression of MMPs is generally low but is increased in the remodeling processes of atherosclerosis and myocardial infarction (Brown et al. 1995; Ducharme et al. 2000). MMPs regulate the degradation of the extracellular matrix and play an important role in cardiac and vascular remodeling (Dollery et al. 1995). Overexpression of MMP-1, -2, -3, -7, -8, -9, -12, -13, -14, and -17 has been observed in atherosclerotic tissues (Carrell et al. 2002). Of these MMPs, MMP-2 and -9 are elevated in the peripheral blood of patients with coronary artery disease and MMP-9 is a predictor of cardiovascular mortality in these patients (Blankenberg et al. 2003). MMP-9 degrades the basement membrane to promote both monocyte infiltration into the plaque and smooth muscle cell migration into the fibrous cap. An increase in MMP-9 activity results in degradation of the fibrous cap, plaque instability, and plaque rupture (Szmitko et al. 2003). Therefore, MMP-9 plays important roles in cardiovascular events. In addition, production of MMP-9 is stimulated by hypoxia and by several cytokines, such as IL-6 and TNF-α (Saren et al. 1996; Kossakowska et al. 1999; Kondo et al. 2002). Although these cytokines are increased and hypoxia is induced by apnea and hypopnea during sleep in patients with OSA, serum levels and activity of MMP-9 and levels of its inhibitor, tissue inhibitor of metalloproteinase-1 (TIMP-1), have not been examined in patients with OSA. Serum levels of MMP-9 and MMP-9 activity were higher in patients with OSA than in obese control subjects but TIMP-1 levels did not differ significantly (Tazaki et al. 2004). In patients with OSA the severity of OSA was the primary factor influencing levels and activity of MMP-9. nCPAP significantly decreased serum levels and activity of MMP-9 but not affect TIMP-1 levels. Therefore, OSA may increase risks of cardiovascular morbidity, and nCPAP might be useful for decreasing these risks.

6 References

Aso Y, Okumura K, Takebayashi K, Wakabayashi S, Inukai T (2003) Relationships of plasma interleukin-18 concentrations to hyperhomocysteinemia and carotid intimal-media wall thickness in patients with type 2 diabetes. Diabetes Care 26:2622-2627

Arzt M, Young T, Finn L, Skatrud JB, Bradley TD (2005) Association of sleep-disordered breathing and the occurrence of stroke. Am J Respir Crit Care Med 172:1447-1451

Barks JL, McQuillan JJ, Iademarco MF (1997) TNF- and IL-4 synergistically increase vascular cell adhesion molecule-1 expression in cultured vascular smooth muscle cells. J Immunol 159:4532-4538

Brown DL, Hibbs MS, Kearney M, Loushin C, Isner JM (1995) Identification of 92-kD gelatinase in human coronary atherosclerotic lesions. Association of active enzyme synthesis with unstable angina. Circulation 91:2125-2131

Blankenberg S, Rupprecht HJ, Poirier O, Bickel C, Smieja M, Hafner G, Meyer J, Cambien F, Tiret L (2003) Plasma concentrations and genetic variation of matrix metalloproteinase 9 and prognosis of patients with cardiovascular disease. Circulation 107:1579-1585

Cao JJ, Thach C, Manolio TA, Psaty BM, Kuller LH, Chaves PH, Polak JF, Sutton-Tyrrell K, Herrington DM, Price TR (2003) C-reactive protein, carotid intima-media thickness, and incidence of ischemic stroke in the elderly: the Cardiovascular Health Study. Circulation 108:166-170

Carrell TW, Burnand KG, Wells GM, Clements JM, Smith A (2002) Stromelysin-1 (matrix metalloproteinase-3) and tissue inhibitor of metalloproteinase-3 are overexpressed in the wall of abdominal aortic aneurysms. Circulation 105:477-482

Dean RT, Wilcox I (1993) Possible atherogenic effects of hypoxia during obstructive sleep apnea. Sleep 16:S15-S22

Dyugovskaya L, Lavie P, Lavie L (2002) Increased adhesion molecules expression and production of reactive oxygen species in leukocytes of sleep apnea patients. Am J Respir Crit Care Med 165:934-939

Dhalla NS, Temsah RM, Netticadan T (2000) Role of oxidative stress in cardiovascular diseases. J Hypertens 18:655-673

Ducharme A, Frantz S, Aikawa M, Rabkin E, Lindsey M, Rohde LE, Schoen FJ, Kelly RA, Werb Z, Libby P, Lee RT (2000) Targeted deletion of matrix metalloproteinase-9 attenuates left ventricular enlargement and collagen accumulation after experimental myocardial infarction. J Clin Invest 106:55-62.

Dollery CM, McEwan JR, Henney AM (1995) Matrix metalloproteinases and cardiovascular disease. Circ Res 77:863-868

Glass CK, Witztum JL (2001) Atherosclerosis: the road ahead. Cell 104:503-516

Haverkate F, Thompson SG, Pyke SD, Gallimore JR, Pepys MB (1997) Production of C-reactive protein and risk of coronary events in stable and unstable angina. European Concerted Action on Thrombosis and Disabilities Angina Pectoris Study Group. Lancet 349:462-466

Hatakeyama H, Nishizawa M, Nakagawa A, Nakano S, Kigoshi T, Uchida K (2002) Testosterone inhibits tumor necrosis factor--induced vascular cell adhesion molecule-1 expression in human aortic endothelial cells. FEBS Lett 530:129-132

Harrison D, Griendling KK, Landmessér U, Hornig B, Drexler H (2003) Role of oxidative stress in atherosclerosis. Am J Cardiol 91:7A-11A

Hodis HN, Mack WJ, LaBree L, Selzer RH, Liu CR, Liu CH, Azen SP (1998) The role of carotid arterial intima-media thickness in predicting clinical coronary events. Ann Intern Med 128:262-269

Kume N, Murase T, Moriwaki H, Aoyama T, Sawamura T, Masaki T, Kita T (1998) Inducible expression of lectin-like oxidized LDL receptor-1 in vascular endothelial cells. Circ Res 83:322-327

Kaynak D, Goksan B, Kaynak H, Degirmenci N, Daglioglu S (2003) Is there a link between the severity of sleep-disordered breathing and atherosclerotic disease of the carotid arteries? Eur J Neurol 10:487-493

Kossakowska AE, Edwards DR, Prusinkiewicz C, Zhang MC, Guo D, Urbanski SJ, Grogan T, Marquez LA, Janowska-Wieczorek A (1999) Interleukin-6 regulation of matrix metalloproteinase (MMP-2 and MMP-9) and tissue inhibitor of metalloproteinase (TIMP-1) expression in malignant non-Hodgkin's lymphomas. Blood 94:2080-2089

Kondo S, Kubota S, Shimo T, Nishida T, Yosimichi G, Eguchi T, Sugahara T, Takigawa M (2002) Connective tissue growth factor increased by hypoxia may initiate angiogenesis in collaboration with matrix metalloproteinases. Carcinogenesis 23:769-776

Lindahl B, Toss H, Siegbahn A, Venge P, Wallentin L (2000) Markers of myocardial damage and inflammation in relation to long-term mortality in unstable coronary artery disease. FRISC Study Group. Fragmin during Instability in Coronary Artery Disease. N Engl J Med 343:1139-1147

Libby P, Ridker PM (1999) Novel inflammatory markers of coronary risk: theory versus practice. Circulation 100:1148-1150

Lavie L (2003) Obstructive sleep apnoea syndrome – oxidative stress disorder. Sleep Med Rev 7:35-51

Lawson JA, Rokach J, FitzGerald GA (1999) Isoprostanes: formation, analysis and use as indices of lipid peroxidation in vivo. J Biol Chem 274:24441-24444

Leonsson M, Hulthe J, Johannsson G, Wiklund O, Wikstrand J, Bengtsson BA, Oscarsson J (2003) Increased Interleukin-6 levels in pituitary-deficient patients are independently related to their carotid intima-media thickness. Clin Endocrinol 59:242-250

Magyar MT, Szikszai Z, Balla J, Valikovics A, Kappelmayer J, Imre S, Balla G, Jeney V, Csiba L, Bereczki D (2003) Early-onset carotid atherosclerosis is associated with increased intima-media thickness and elevated serum levels of inflammatory markers. Stroke 34:58-63

Minoguchi K, Tazaki T, Yokoe K, Minoguchi H, Watanabe Y, Yamamoto M, Adachi M (2004) Elevated production of tumor necrosis factor-a by monocytes in patients with obstructive sleep apnea syndrome. Chest 126:1473-1479

Minoguchi K, Yokoe T, Tanaka A, Ohta S, Hirano T, Yoshino G, O'donnell CP, Adachi M (2006) Association between lipid peroxidation and inflammation in obstructive sleep apnoea. Eur Respir J 28:378-385

Minoguchi K, Yokoe K, Tazaki T, Minoguchi H, Tanaka A, Oda N, Okada S, Ohta S. Naito H, Adachi M (2005) Increased carotid intima-media thickness and serum inflammatory markers in obstructive sleep apnea. Am J Respir Crit Care Med 172:625-630

Minoguchi K, Yokoe T, Tazaki T, Minoguchi H, Oda N, Tanaka A, Yamamoto M, Ohta S, O'Donnell CP, Adachi M (2007) Silent brain infarction and platelet activation in obstructive sleep apnea. Am J Respir Crit Care Med 175:612-617

O'Leary DH, Polak JF (2002) Intima-media thickness: a tool for atherosclerosis imaging and event prediction. Am J Cardiol 90:18-21

O'Leary DH, Polak JF, Kronmal RA, Manolio TA, Burke GL, Wolfson SK Jr (1999) Carotid-artery intima and media thickness as a risk factor for myocardial infarction and stroke in older adults. Cardiovascular Health Study Collaborative Research Group. N Engl J Med 340:14-22

Pasceri V, Willerson JT, Yeh ET (2000) Direct proinflammatory effect of C-reactive protein on human endothelial cells. Circulation 102:2165-2168

Pasceri V, Cheng JS, Willerson JT, Yeh ET (2001) Modulation of C-reactive protein-mediated monocyte chemoattractant protein-1 induction in human endothelial cells by anti-atherosclerosis drugs. Circulation 103:2531-2534

Ridker PM, Rifai N, Stampfer MJ, Hennekens CH (2000) Plasma concentration of interleukin-6 and the risk of future myocardial infarction among apparently healthy men. Circulation 101:1767-1772

Schulz R, Mahmoudi S, Hattar K, Sibelius U, Olschewski H, Mayer K, Seeger W, Grimminger F (2000) Enhanced release of superoxide from polymorphonuclear neutrophils in obstructive sleep apnea. Impact of continuous positive airway pressure therapy. Am J Respir Crit Care Med 162:566-570

Silvestrini M, Rizzato B, Placidi F, Baruffaldi R, Bianconi A, Diomedi M (2002) Carotid artery wall thickness in patients with obstructive sleep apnea syndrome. Stroke 33:1782-1785

Suzuki T, Nakano H, Maekawa J, Okamoto Y, Ohnishi Y, Yamauchi M, Kimura H (2004) Obstructive sleep apnea and carotid-artery intima-media thickness. Sleep 27:129-133

Schulz R, Seeger W, Fegbeutel C, Husken H, Bodeker RH, Tillmanns H, Grebe M (2005) Changes in extracranial arteries in obstructive sleep apnea. Eur Respir J 25:69-74

Szmitko PE, Wang CH, Weisel RD, Jeffries GA, Anderson TJ, Verma S (2003) Biomarkers of Vascular Disease Linking Inflammation to Endothelial Activation: Part II. Circulation 108:2041-2048

Saren P, Welgus HG, Kovanen PT (1996) TNF-alpha and IL-1beta selectively induce expression of 92-kDa gelatinase by human macrophages. J Immunol 157:4159-4165

Tazaki T, Minoguchi K, Yokoe T, Samson KT, Minoguchi H, Tanaka A, Watanabe Y, Adachi M (2004) Increased levels and activity of matrix metalloproteinase-9 in obstructive sleep apnea syndrome. Am J Respir Crit Care Med 170:1354-1359

Visse R, Nagase H (2003) Matrix metalloproteinases and tissue inhibitors of metalloproteinases: structure, function, and biochemistry. Circ Res 92:827-839

Wang TJ, Nam BH, Wilson PW, Wolf PA, Levy D, Polak JF, D'Agostino RB, O'Donnell CJ (2002) Association of C-reactive protein with carotid atherosclerosis in men and women: the Framingham Heart Study. Arterioscler Thromb Vasc Biol 22:1662-1667

Yudkin JS, Kumari M, Humphries SE, Mohamed-Ali V (2000) Inflammation, obesity, stress and coronary heart disease: is interleukin-6 the link? Atherosclerosis 148:209–214

Yokoe T, Minoguchi K, Matsuo H, Oda N, Minoguchi H, Yoshino G, Hirano T, Adachi M (2003) Elevated levels of C-reactive protein and interleukin-6 in patients with obstructive sleep apnea syndrome are decreased by nasal continuous positive airway pressure. Circulation 107:1129-1134

Yaggi HK, Concato J, Kernan WN, Lichtman JH, Brass LM, Mohsenin V (2005) Obstructive sleep apnea as a risk factor for stroke and death. N Engl J Med 353:2034-2041

Significance of Small Dense Low-Density Lipoproteins in Coronary Heart Disease

Shinji Koba[1], Tsutomu Hirano[2], Yuuya Yokota[1], Fumiyoshi Tsunoda[1], Yoshihisa Ban[1], Takayuki Sato[1], Makoto Shoji[1], Hiroshi Suzuki[1], Eiichi Geshi[1], Takashi Katagiri

[1]Third Department of Internal Medicine and [2]First Department of Internal Medicine, Showa University School of Medicine, 1-5-8 Hatanodai, Shinagawa-ku, Tokyo 142-8666, Japan

Summary. Low-density lipoprotein (LDL) particles are heterogeneous with respect to their size, density, and lipid composition, and the size of LDL particles is chiefly determined by their lipid contents. Small dense LDL particles have been suggested to be highly atherogenic compared to large buoyant LDL. Our case-control studies have shown that the LDL particle size determined by gradient gel electrophoresis was remarkably smaller in patients with coronary heart disease (CHD), irrespective of the presence of diabetes and the differences in clinical situation and severity of CHD. In addition, small dense LDL-cholesterol concentration evaluated by heparin magnesium precipitation was significantly higher in severe stable CHD and acute coronary syndrome compared with non-CHD subjects and patients with mild CHD, while large LDL-cholesterol estimated by subtracting the small dense LDL-cholesterol concentration from the LDL-cholesterol concentration, were somewhat lower in stable CHD compared with healthy subjects. Furthermore, reduced LDL particle size and elevated small dense LDL-cholesterol levels were significantly associated with metabolic dyslipidemia in Metabolic syndrome. These suggest that the predominance of small dense LDL and high levels of small dense LDL-cholesterol are very promising risk marker for CHD.

Key words. small dense LDL, coronary heart disease, small dense LDL-cholesterol, metabolic syndrome

1 Introduction

Low-density lipoprotein (LDL) particles contain one large molecule of apolipoprotein (apo) B as their constituent protein and cholesterol as the major lipid component. LDL particles are heterogeneous with respect to their size, density, and lipid composition, and the size of LDL particles is chiefly determined by their lipid contents. High plasma levels of LDL-cholesterol (LDL-C) is an established major risk factor for coronary heart disease (CHD). However, previous studies (Genest Jr et al. 1992, Kannel et al. 1995) have shown that the difference of distribution of LDL-C is very small between CHD patients who did not take any lipid-lowering drugs and non-diseased population, and about 80% of patients had LDL-C levels in the same range compared to controls. On the other hand, the distribution of high-density lipoprotein (HDL)-C concentration shift toward the lower levels about 10 mg/dL and the distribution of apo B towards the higher levels about 10 mg/dL compared to the controls. More than 2 decades ago, Sniderman and his colleagues (Sniderman et al. 1980) compared the cholesterol content and apo B content in LDL fractions separated by ultracentrifugation in 31 patients without and 59 patients with angiographically documented CHD. They showed that not LDL-C content but the LDL-apo B content was the better marker to discriminate between patients with and without CHD. Thus, HDL-C or apo B appears to be a better marker to discriminate CHD than LDL-C. Figure 1 shows the same LDL-C but different LDL particle numbers in two subjects. The subject A has a predominance of cholesterol-rich large LDL particles whereas the subject B has many cholesterol-poor small LDL particles, resulting in marked difference in small LDL-C concentration. These changes in the composition of LDL particles are associated with the changes of triglyceride and apo B levels. Subject B is often seen in CHD and/or Metabolic syndrome.

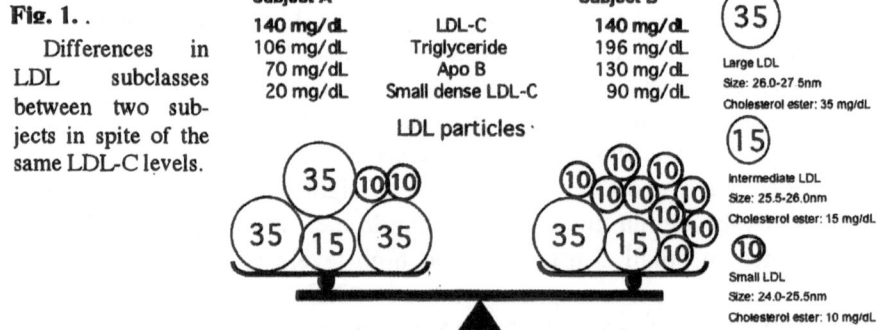

Fig. 1..
Differences in LDL subclasses between two subjects in spite of the same LDL-C levels.

Subject A		Subject B
140 mg/dL	LDL-C	140 mg/dL
106 mg/dL	Triglyceride	196 mg/dL
70 mg/dL	Apo B	130 mg/dL
20 mg/dL	Small dense LDL-C	90 mg/dL

LDL particles

Large LDL
Size: 26.0-27.5nm
Cholesterol ester: 35 mg/dL

Intermediate LDL
Size: 25.5-26.0nm
Cholesterol ester: 15 mg/dL

Small LDL
Size: 24.0-25.5nm
Cholesterol ester: 10 mg/dL

2 Why is Small Dense LDL Atherogenic?

Chapman's group (Chancharme et al. 1999, Goulinet et al. 1997) have shown that small dense LDLs (sd-LDLs) separated by density-gradient ultracentrifugation, contain diminished amounts of tocopherols and carotenoids, lipophilic antioxidants, and contain less amounts of polyunsaturated molecular species of phosphatidylcholine and cholesteryl esters. These suggest that sd-LDL display increased susceptibility to oxidation. In addition, sd-LDL particles have a lower binding affinity for the LDL receptor and have longer clearance time compared with large LDL. Sd-LDL penetrates into the arterial wall much more and faster than native LDL particle. Further more, sd-LDL and oxidized LDL bind to extracellular matrix more tightly than native LDL. These all results in deposition of cholesterol in the arterial wall and progression of atherosclerosis (Berneis et al. 2002).

Fig. 2. Comparison of LDL particle size and small dense LDL phenotype between non-diabetic men with CHD and age-matched healthy control. *p<0.05 vs control. Based on reference Koba et al. 2000.

	Control (N=42)	CHD (N=87)
Age	61±11	60±11
Body mass index	23.0±2.9	23.6±3.1
Smoker (%)	60	56
LDL-cholesterol (mg/dL)	116±33	119±26
HDL-cholesterol (mg/dL)	49.8±14.0	42.7±10.1*
Triglyceride (mg/dL)	119±64	137±77
Apolipoprotein B (mg/dL)	90±24	104±24*
Remnant-like particle cholesterol (mg/dL)	5.2±2.6	6.5±5.4
Insulin (μU/mL)	6.3±4.1	6.9±3.9

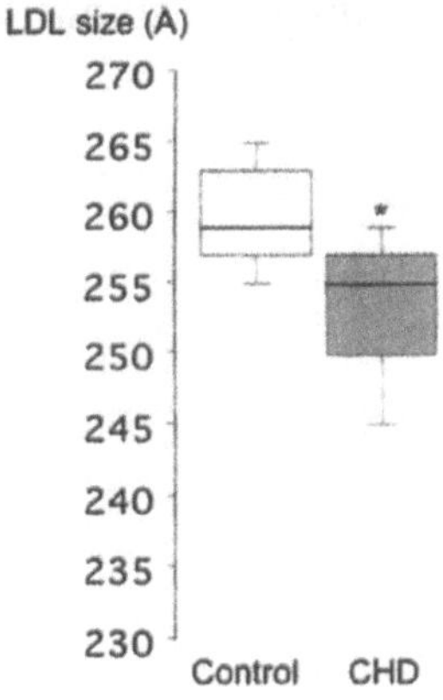

LDL size (A) — box plot comparison of Control and CHD (axis 230–270)

3 Small Dense LDL and CHD

3.1 Measurement of LDL particle size

LDL particle size is most often measured by gradient gel electrophoresis using non-denatured 2 to 16% polyacrylamide gel according to the procedure described by Nichols et al (Nichols et al. 1986). Two distinct LDL size phenotypes, pattern A, large buoyant LDL particles and pattern B, sd-

LDL particles, can be easily separated. We compared the LDL size and prevalence of sd-LDL by gradient gel electrophoresis in 87 non-diabetic patients with CHD and in 42 age-matched healthy men (Koba et al. 2000). HDL-C was significantly lower and apo B was significantly higher in CHD patients, however, no significant differences were seen in LDL-C, triglyceride, remnant-like particle cholesterol, and insulin levels. The mean LDL particle size was remarkably smaller in CHD patients and three fourths of patients showed pattern B (LDL size ≤ 25.5nm) (Figure 2). Stepwise regression analysis revealed that LDL size was the most powerful independent determinant of CHD (F value 15.3; p<0.001). We then measured LDL size in 571 patients with CHD and in 263 healthy men and women (Koba et al. 2002). The LDL size was significantly smaller in CHD patients in both men and women, and the incidence of pattern B was significantly higher in any types of CHD such as vasospastic angina, stable CHD and acute coronary syndrome (ACS). These changes in LDL size and prevalence of pattern B were constant when all diabetic patients and patients on lipid-lowering therapy were excluded (Figure 3). However, the LDL size was comparable irrespective of the type of CHD and the extent and severity of the coronary lesions.

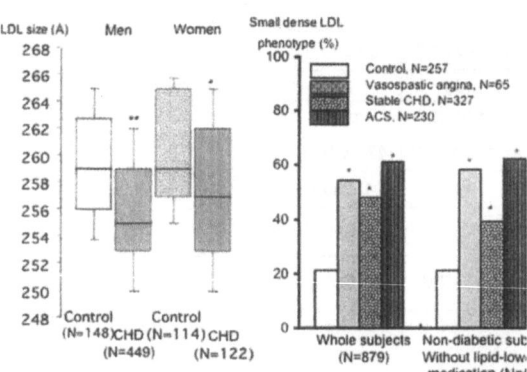

Fig. 3. Comparison of LDL particle size between CHD and healthy subjects in both men and women, and prevalence of small dense LDL phenotype among control, vasospastic angina, stable CHD, and ACS. *p<0.05 **P<0.0001 vs control. Based on reference Koba et al. 2002.

3.2 Measurement of Small Dense LDL-Cholesterol

A recent report from the Québec cardiovascular study including 4635 French Canadian men without any cardiovascular disease has confirmed that a higher amounts of small LDL-C at baseline, determined by semi-quantitative analysis of cholesterol content in both large and small LDL subclasses by the gradient gel electrophoresis, is a strong and independent predictor of first onset of CHD in the first 7 years of follow-up whereas elevated large LDL-C is not predictors of CHD; rather have a somewhat favorable CHD risk (St-Pierre et al. 2005). This suggests that the atherogenicity of LDL differ among the heterogeneity of LDL particles and majority of the association of atherogenic LDL-C with CHD is due to sd-LDL component. However, gradient gel electrophoresis does not allow quantitative determination of sd-LDL-C.

Hirano and co-workers have recently discovered a simple and rapid method for measuring sd-LDL-C by heparin magnesium precipitation (Hirano et al. 2005). Briefly, the precipitation reagent containing heparin and magnesium is added to each serum sample followed by incubation, then the samples are centrifuged and the pass through fraction is collected for the measurement. The clear infranatant is then analysed by direct homogenous LDL-C methodologies. That is sd-LDL-C concentration.

Mean ± SE

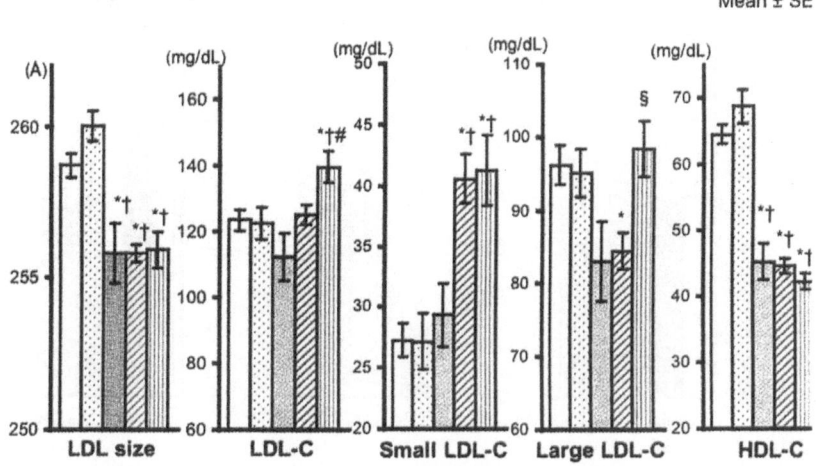

Fig. 4. Comparison of LDL size, LDL-C, small LDL-C, large LDL-C, and HDL-C among healthy men (N=95), healthy postmenopausal women (N=47), vasospastic angina (N=18), stable CHD (N=123), and ACS (N=84). *$p<0.05$ vs control men, †$p<0.05$ vs control women, #$p<0.05$ vs vasospastic angina, §$p<0.05$ vs stable CHD. Based on reference Koba et al. 2006.

We compared LDL size and sd-LDL-C concentration in 225 consecutive angiographically documented CHD patients who were not receiving any lipid-lowering medication, and 95 healthy men, aged 40 to 63 years and 47 healthy postmenopausal women (Koba et al. 2006). Similar to our previous study (Koba et al. 2002), LDL particle size and HDL-C levels were significantly lower in all types of CHD compared with healthy men and women, while LDL-C levels were significantly higher only in ACS patients. Sd-LDL-C levels were significantly higher in both stable CHD and ACS. On the other hand, large LDL-C, estimated by subtracting the sd-LDL-C concentration from the LDL-C concentration, was somewhat lower in vasospastic angina and stable CHD (Figure 4). We determine the significance of sd-LDL-C concentration with respect to severity of coronary atherosclerosis (Figure 5). The severity of coronary atherosclerosis was estimated by calculating the Gensini score by coronary arteriography findings. Sd-LDL-C gradually increased as the Gensini score increased. Large LDL-C did not differ among the different quartile of Gensini score, and were somewhat lower compared with healthy men and women. This trend of increased sd-LDL-C along with the increased Gensini score was even more pronounced when the diabetic patients were excluded.

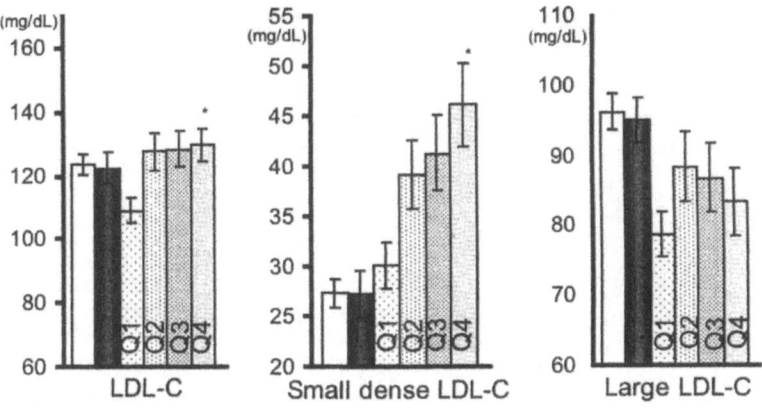

Fig. 5. Associations of LDL-C levels, small dense LDL-C levels, and large LDL-C levels with the CHD severity (Quartile of Gensini score). Quartile 1 (Q1) = Gensini score ≤ 7.5; quartile 2 (Q2) = 9.0-26.0; quartile 3 (Q3) = 27.0-49.0; quartile 4 (Q4) = the score ≥ 52.0. The overall population included 36, 36, 34, and 35 patients in Q1, Q2, Q3, and Q4, respectively. *p<0.05 vs Q1, † p<0.05 vs Q2 by by Tukey-Kramer post-hoc test. Based on reference Koba et al. 2006.

[   ] Healthy men   � healthy postmenopausal women

4 Small Dense LDL-Cholesterol and Metabolic Syndrome

It is well-known that LDL size is positively associated with HDL-C and is negatively associated with triglyceride levels. These lipid levels are two of the major components of diagnosis in Metabolic syndrome. We compared LDL particle size, LDL-C, sd-LDL-C, and large LDL-C among normolipidemic individuals, individuals with either high triglyceride (≥150 mg/dL) or low HDL-C (<40 mg/dL), and individuals with metabolic dyslipidemia (normal LDL-cholesterolemic [<140 mg/dL] subjects with high triglyceride and low HDL-C) among 389 un-medicated subjects. The LDL particle size was significantly smaller, and sd-LDL-C levels were more than doubled whereas large LDL-C levels were significantly lower in patients with metabolic dyslipidemia, in spite of similar LDL-C levels (Figure 6). This suggests that the sd-LDL particles and high concentrations of sd-LDL-C concentration are strongly associated with metabolic dyslipidemia. Similar findings of LDL subclasses are shown by nuclear magnetic resonance (NMR) analysis of lipoproteins in the Recent Framingham Heart Study, which included around 3000 men and women without cardiovascular disease (Kathiresan et al. 2006). In that study, plasma triglyceride gradually increased and HDL-C gradually decreased and NMR-determined small LDL particle number increased whereas large LDL particle number decreased along with the increased number of components of Metabolic syndrome in both men and women (Figure 7).

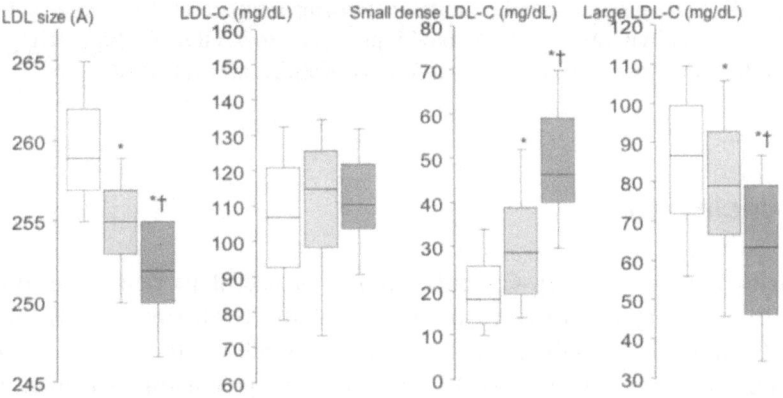

Fig. 6. Associations of LDL size, LDL-C, sd-LDL-C, and large LDL-C with metabolic dyslipidemia (i.e., high triglyceride and/or low HDL-C) in nomolipidemic subjects (171 male and 97 female), in individuals with either high triglyceride (≥150 mg/dL) or low HDL-C (<40 mg/dL) (73 male and 16 female), and in individuals with both high triglyceride and low HDL-C (24 male and 8 female). The box encompasses the 25th through 75th percentiles, and the 5th and 95th percentiles are shown as horizontal lines. The LDL particle size, sd-LDL-C levels, and large-LDL-C levels were significantly smaller, significantly higher, and significantly lower, respectively, in individuals with either high triglyceride than in nomolipidemic subjects. These same changes, meanwhile, were even further pronounced in individuals with both high triglyceride and low HDL-C in spite of similar LDL-C levels. *p<0.05 vs normolipidemic subjects, †p<0.05 vs subjects with either high triglyceride or low HDL-C.

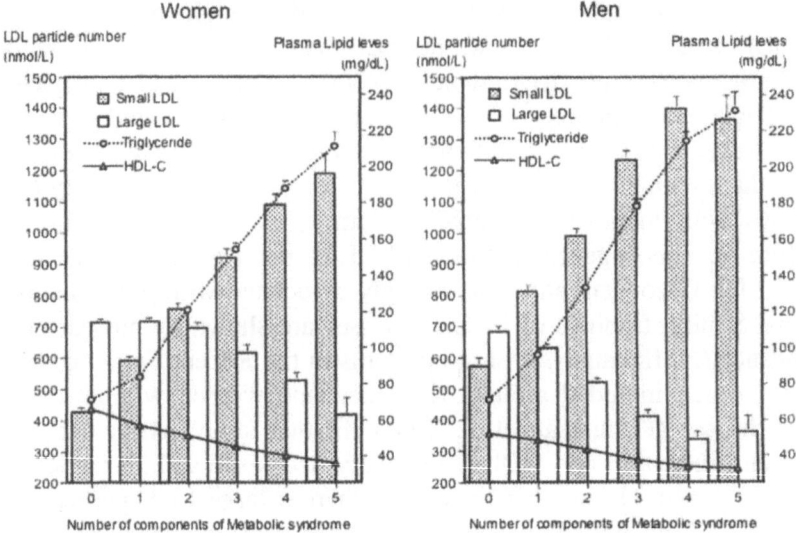

Fig. 7. Plasma levels of triglyceride and HDL-C, and NMR-determined large LDL and small LDL particle number with increasing number of Metabolic syndrome feature (abdominal obesity, high blood pressure, low HDL-C, high triglyceride, and high fasting glucose). Based on reference Kathiresan et al. 2006.

5 Conclusion

Measurement of sd-LDL-C concentration is useful to detect the progression of coronary atherosclerosis, to predict the cardiovascular events, and to determine the metabolic dyslipidemia. Therefore, the predominance of sd-LDL and high levels of sd-LDL-C are very promising risk markers for CHD.

References

Berneis KK, Krauss RM (2002) Metabolic origins and clinical significance of LDL heterogeneity. J Lipid Res 43:1363-79.

Chancharme L, Therond P, Nigon F, Lepage S, Couturier M, Chapman MJ (1999) Cholesteryl ester hydroperoxide lability is a key feature of the oxidative susceptibility of small, dense LDL. Arterioscler Thromb Vasc Biol 19:810-820.

Genest Jr J, McNamara JR, Ordovas JM, Jenner JL, Silberman SR, Anderson KM, Wilson PF, Salem DN, Schaefer EJ (1992) Lipoprotein cholesterol, apolipoprotein A-I and B and Lipoprotein (a) abnormalities in men with premature coronary artery disease. J Am Coll Cardiol 19:792-802.

Goulinet S, Chapman MJ (1997) Plasma LDL and HDL subspecies are heterogenous in particle cpntent of tocopherols and oxygenated and hydrocarbon carotenoids. Relevance to oxidative resistance and atherosclerosis. Arterioscler Thromb Vasc Biol 17:786-796.

Hirano T, Ito Y, Yoshino G (2005) Measurement of small dense low-density lipoprotein particles. J Atheroscler Thromb 12:67-72.

Kannel WB (1995) Range of serum cholesterol values in the population developing coronary artery disease. Am J Cardiol 76:69C-77C.

Kathiresan S, Otovas JD, Sullivan LM, Keyes MJ, Schaefer EJ, Wilson PWF, D'Agostino RB, Vasan RS, Robins SJ (2006) Increased small low-density lipoprotein particle number. A prominent feature of the Metabolic syndrome in the Framingham Heart Study. Circulation 113:20-29.

Koba S, Hirano T (2000) Small dense low-density lipoprotein in Japanese men with coronary artery disease. Ann Intern Med 132:762.

Koba S, Hirano T, Kondo T, Shibata M, Suzuki H, Murakami M, Geshi E, Katagiri T (2002) Significance of small dense low-density lipoproteins and other risk factors in patients with various types of coronary heart diseases. Am Heart J 144:1026-35.

Koba S, Hirano T, ItoY, Tsunoda, T, Yokota Y, Ban Y, Iso Y, Suzuki H, Katagiri T (2006) Significance of small dense low-density lipoprotein-cholesterol concentrations in relation to the severity of coronary heart diseases. Atherosclerosis 189:206-214.

Nichols AV, Krauss RM, Musliner TA (1986) Nondenaturing polyacrylamide gradient gel electrophoresis. Methods Enzymol 128:417-433.

Sniderman A, Shapiro S, marpole D, Skinner B, Teng B, Kwiterovich Jr PO (1980) Association of coronary atherosclerosis with hyper*apo*betalipoproteina [increased protein but normal cholesterol levels in human plasma low density (ß) lipoproteins]. Proc Natl Acad Sci USA 77:604-608.

St-Pierre AC, Cantin B, Dagenais GR, Mauriége P, Bernard PM, Després JP, Lamarche B (2005) Low-density lipoprotein subfractions and the long-term risk of ischemic heart disease in men, 13-year follow-up data from the Québec cardiovascular study. Arterioscler Thromb Vasc Biol 25:553-559.

Statin decreases IL-1 and LPS-induced inflammatory cytokines production in oral epithelial cells

Michihiko Usui[1], Reiko Suda[1], Yasushi Miyazawa[1], Makoto Kobayashi[1], Yoshimasa Okamatsu[1,2], Hisashi Takiguchi[1], Motoyuki Suzuki[1] and Matsuo Yamamoto[1]

[1]Department of Periodontology, Showa University School of Dentistry, 2-1-1 Kitasenzoku, Ohta-ku, Tokyo 145-8515, Japan
[2]Dental Clinic, Showa University Hospital, 1-5-8 Hatanodai, Shinagawa-ku, Tokyo 142-8666, Japan

<e-mail>yamamoto@senzoku.showa-u.ac.jp

Summary. Periodontitis is a choronic inflammatory disease associated with degradation of periodontal tissues. This disease is considered to result of production of proinflammatory cytokines and tissue degradative enzymes which are initiated and advanced by lipopolysaccharide (LPS) of oral bacteria.

Statin, 3-hydroxy-3-mrthylglutaryl coenzyme A (HMG-CoA) reductase inhibitor, is known as a drug lowering serum cholesterol. In hyperlipidemia, the increase of cholesterol develops atherosclerosis in blood vessel. Many studies revealed that statins also have the cholesterol-independent, also termed pleiotropic, effects on variety of vascular cells, and decrease the expression level of inflammatory cytokines such as interleukin (IL)-1, IL-6 and IL-8.

As a chronic inflammatory disease, periodontitis shares similar mechanisms with atherosclerosis. We measured the effect of simvastatin, one of statins, on IL-1 and LPS-induced IL-6 and IL-8 production in oral epithelial cells. Simvastatin decreased expression of these inflammatory cytokines and promoter activity of downstream pathway. Our results suggested that statin might be helpful tool for periodontitis.

Key words. Statin (3-hydroxy-3-mrthylglutaryl coenzyme A reductase inhibitor), inflammatory cytokines, periodontitis, oral epithelial cells

Introduction

Periodontitis is inflammatory degradation disease in periodontal tissue, which is composing of gingiva, cementum, periodontal ligament, and alveolar bone. Periodontal disease is considered to result from production of proinflammatory cytokines and tissue degradative enzymes which are initiated and advanced by oral bacterial infection (Okada and Murakami 1998). *Porphyromonas gingivalis (P.g.)*, a Gram-negative, anaerobic bacterium, has been implicated in the etiology and pathogenesis of chronic periodontitis. This microorgamism is isolated from periodontal pocket of periodontal patient frequently. The LPS of *P.g.* is a potent stimulator of inflammatory cytokine production and tissue destruction in periodontitis, resulting in tooth loss. Recently, this *P.g.* was reported to associate with cardiovascular disease progression (Scannapieco et al. 2003). On the other hands, Interleukin (IL)-1 is a multipotential cytokine that mediates inflammatory tissue destruction such as periodontal tissue in periodontitis (De Nardin 2001).

Statins such as simvastatin are pharmacologic 3-hydroxy-3-mrthylglutaryl coenzyme A (HMG-CoA) reductase inhibitors constitute the most powerful class of lipid lowering drugs. Clinical study has demonstrated a marked reduction in cardiovasucular mortality in patients taking statins (O'Driscoll et al. 1997). The benefits observed with statin treatement appear to be related their pleiotropic effect independent of cholesterol-lowering (Takemoto et al. 2001). Furthermore, many clinical trials of statin in last decade showed that the clinical benefits of statin could be related to anti-inflammatory properties, an improvement in endothelial dysfunction, a reduction in blood thrombogenicity and immunomodulatory actions. Recently, these pleiotropic effects were caused of inhibition of small G proteins prenylation by statins in atherogenic process (Takemoto et al. 2001).

Here we investigated the effect of statin on IL-1 and LPS induced proinflmantory cytokine in epithelial cells.

Statin decreases IL-1-induced proinflammatory cytokines in oral epithelial cells.

To determine if statin such as simvastatin affects on IL-1 induced proinflammatory cytokines, which are IL-6 and IL-8 in epithelial cells, KB cells were treated with IL-1 (1ng/ml) and simvastatin (10^{-8}-10^{-5}M). IL-1 increased IL-6 and IL-8 production in KB cells as expected. Simvastatin treatment decreased IL-1 induced IL-6 and IL-8 production dose-dependently (Fig. 1A). Furthermore, 10^{-6}M simvastatin administration also suppressed IL-1 induced IL-6 and Il-8 production in primary human gingival fibroblasts (HGF) (Fig. 1B).

Statin also decreases LPS-induced inflammatory cytokine in oral epithelial cells.

LPS from *P.g.* is known to be a stimulator of cytokine production in periodontal tissues. Recently, it is demonstrated that CD14 and Toll-like Receptor 4 are expressed on the surface of HGFs (Gutierrez-Venegas et al. 2006). The dimerized these molecules recognize LPS, and transducer the signal into nuclei through MAPK pathway in HGFs, similar to typical response in macrophage. LPS treatment stimulated IL-6 production in HGFs similar to IL-1. To determine whether simvastatin decreases LPS-induced IL-6 production, HGFs were treated with LPS (1ug/ml) and simvastatin (10^{-6}M). Simvastatin administration decreased LPS-induced IL-6 production in HGFs similar to KB cells (Fig. 2).

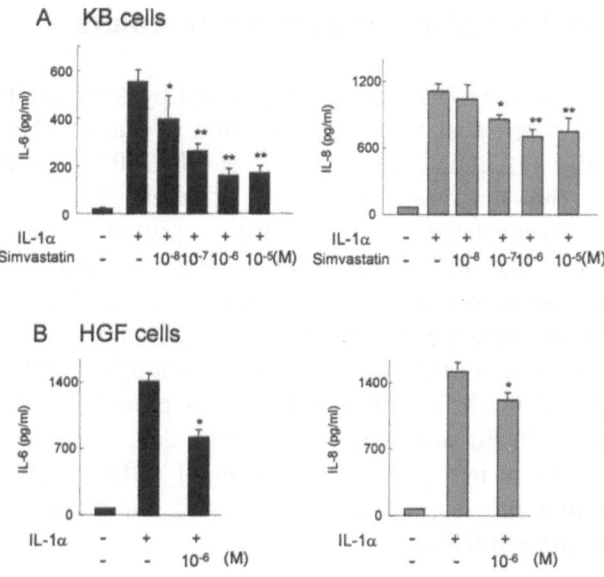

Fig. 1. Effects of simvastatin on IL-1 induced IL-6 and 8 production in epithelial cells. Cellular expression of IL-6 and IL-8 is shown in cultures after treatment

with the indicated concentrations of simvastatin for 7 hr, and/or IL-1 (1 ng/ml) for the last 5 of these hours in KB cells (A) and human gingival fibroblasts (HGFs) (B). Data are expressed as means ± SD (n = 4). * $P < 0.01$ and ** $P < 0.001$ vs. control (treated with IL-1)

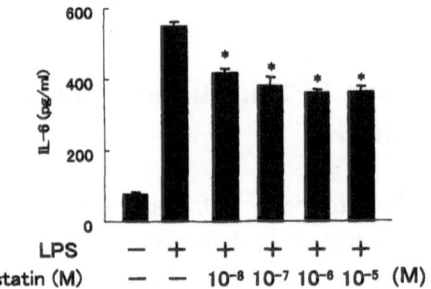

Fig. 2. Dose-dependent effects of simvastatin on LPS induced IL-6 production in HGFs. Cellular expression of IL-6 is shown in cultures after treatment with the indicated concentrations of simvastatin for 7 hr, and/or LPS of *P.g.* (1 ug/ml) for the last 5 of these hours in HGFs cells. Data are expressed as means ± SD (n = 4). * $P < 0.01$ and ** $P < 0.001$ vs. control (treated with LPS)

Statin has pleiotropic effects on epithelial cells through regulatory function of isoprenoid intermediate.

Farnesyl pyrophosphate (FPP) and geranylgeranyl pyrophosphate (GGPP) are important for post-translational modification of small GTPase of the Ras / Rho family (Etienne-Manneville S et al. 2002). Prenylation is prerequisite for the activation of these proteins. Ras proteins are predominantly farnesylated, while Rho proteins are mainly geranylgeranylated. To investigate whether Ras or Rho proteins are involved in the simvastatin mediated reduction of IL-6 and IL-8 production, KB cells were incubated with simvastatin in the presence of an isoprenoid intermediate, either FPP or GGPP. Although GGPP prevented simvastatin-imduced inhibition of IL-6 and IL-8 production in KB cells treated by IL-1, FPP did not (Fig. 3A). These data suggested that the mevalonate pathway is involved in regulation of inflammatory cytokine expression and that the isoprenylation of Rho is involved in the IL-1 induced inflammation in KB cells.

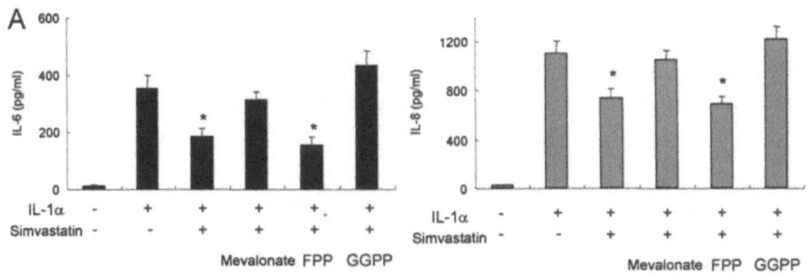

B

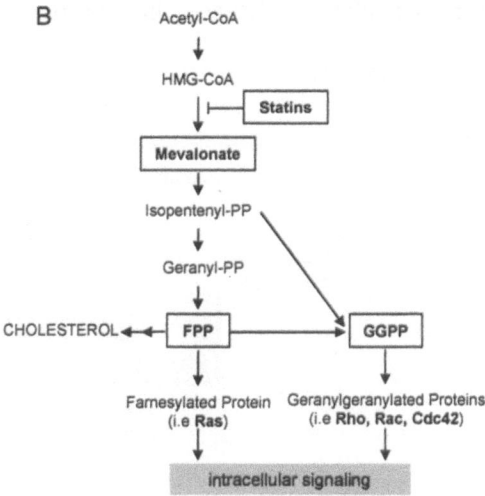

Acetyl-CoA

HMG-CoA

Statins

Mevalonate

Isopentenyl-PP

Geranyl-PP

CHOLESTEROL ← FPP → GGPP

Farnesylated Protein   Geranylgeranylated Proteins
(i.e **Ras**)   (i.e **Rho, Rac, Cdc42**)

intracellular signaling

Fig. 3. Reversal of inhibitory effect of simvastatin on epthelial cells by co-treatment with downstream metabolites of HMG-CoA reductase.(A) Inhibitory effects of simvastatin (10^{-6} M) on KB cells were reversed by co-treatment with mevalonate (10^{-4}M) or GGPP (5×10^{-6}M), but not with FPP (5×10^{-6}M). Data are expressed as means ± SD (n = 4).   * $P < 0.001$ vs. control (treated with IL-1) (B)Schematic representation of the mevalonate pathway. Statins block conversion of HMG-CoA to mevalonate. This leads to reduced synthesis of cholesterol and decreased prenylation of proteins such as small GTPases. Isopentenyl-PP, isopentenyl pyrophosphate; Geranyl-PP, geranyl pyrophosphate.

Statin suppresses IL-1 induced AP-1 and NF-kB promoter activity.

AP-1 and NF-kB are known to be downstream transcriptional factors for IL-1 and LPS signaling (Jung et al. 2002, Mackman et al. 1991). We examined whether simvastatin effect IL-1 induced AP-1 and NF-kB promoter activity in KB cells. IL-1 stimulated AP-1 and NF-kB promoter activity reported previously (Jung et al. 2002). Simvastatin treatment suppresses IL-1 induced AP-1 and NF-kB activity (Fig. 4). NF-kB and AP-1 coordinate the expression of a wide variety of genes that regulate immune responses, and are involved in many inflammatory diseases, statin might be a beneficial tool for periodontal disease in addition to atherosclerosis.

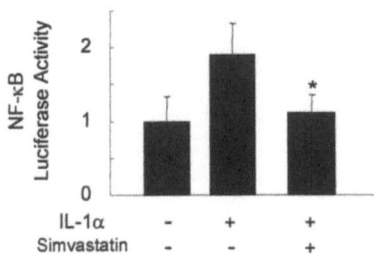 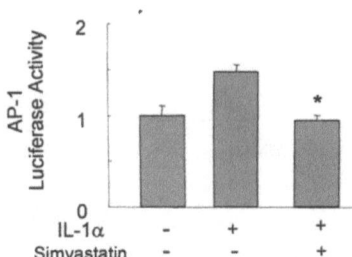

Fig. 4. Suppression of NF-kB and AP-1 activity in IL-1-stimulated KB cells by simvastatin. KB cells were transiently cotransfected with pNF-kB-luc or pAP-1-luc, together with pCMV-βgal as a control. The cells were analyzed 48 hr later, with 5 hr stimulation with IL-1 (1 ng/ml). All results were normalized for transfection efficiency using expression of β-galactosidase. Data are expressed as means ± SD (n = 4). * $P < 0.001$ vs. control (treated with IL-1)

Periodontitis and Statins

Periodontitis has been known to be a risk factor for cardiovascular disease due to elevated inflammatory cytokines such as IL-1 and TNF-α in periodontal lesions and consequently increased serum levels (Beck and Offenbacher 2001). It is possible that administration of statins to cardiovascular patients may have additional benefit on suppressing inflammation in periodontal tissue. In this review, we showed statin suppresses IL-1 and LPS from *P.g.* induced IL-6 and IL-8 production in endothelial cells in vitro.

It is reported that statin use decreased the rate of tooth loss in chornic peridontitis patients (Cunha-Cruz et al. 2006). In addition, Mundy demonstrated that statins increases new bone formation in vitro and in rodents (Mundy et al. 1999). These reports and our findings suggested that statins may be a helpful tool for periodontitis.

References

Beck JD, Offenbacher S (2001) The association between periodontal diseases and cardiovascular diseases: a state-of-the-science review. Ann Periodontol 6:9-

15.

Cunha-Cruz J, Saver B, Maupome G, Hujoel PP (2006) Statin use and tooth loss in chronic periodontitis patients. J Periodontol 77:1061-1066

De Nardin E (2001) The role of inflammatory and immunological mediators in periodontitis and cardiovascular disease. Ann Periodontol 6:30-40

Etienne-Manneville S, Hall A (2002) Rho GTPases in cell biology. Nature 420:629-635.

Gutierrez-Venegas G, Kawasaki-Cardenas P, Cruz-Arroyo SR, Perez-Garzon M, Maldonado-Frias S (2006) Actinobacillus actinomycetemcomitans lipopolysaccharide stimulates the phosphorylation of p44 and p42 MAP kinases through CD14 and TLR-4 receptor activation in human gingival fibroblasts. Life Sci. 78:2577-2583.

Jung YD, Fan F, McConkey DJ, Jean ME, Liu W, Reinmuth N, Stoeltzing O, Ahmad SA, Parikh AA, Mukaida N, Ellis LM (2002) Role of P38 MAPK, AP-1, and NF-kappaB in interleukin-1beta-induced IL-8 expression in human vascular smooth muscle cells. Cytokine 18:206-213.

Mackman N, Brand K, Edgington TS (1991) Lipopolysaccharide-mediated transcriptional activation of the human tissue factor gene in THP-1 monocytic cells requires both activator protein 1 and nuclear factor kappa B binding sites. J Exp Med. 174:1517-1526.

Mundy G, Garrett R, Harris S, Chan J, Chen D, Rossini G, Boyce B, Zhao M, Gutierrez G (1999) Stimulation of bone formation in vitro and in rodents by statins. Science 286:1946-1949

O'Driscoll G, Green D, Taylor RR (1997) Simvastatin, an HMG-coenzyme A reductase inhibitor, improves endothelial function within 1 month. Circulation 95:1126-1231.

Okada H, Murakami S (1998) Cytokine expression in periodontal health and disease. Crit Rev Oral Biol Med 9:248-266

Scannapieco FA, Bush RB, Paju S (2003) Associations between periodontal disease and risk for atherosclerosis, cardiovascular disease, and stroke. A systematic review. Ann Periodontol 8:38-53

Takemoto M, Liao JK (2001) Pleiotropic effects of 3-hydroxy-3-methylglutaryl coenzyme a reductase inhibitors. Arterioscler Thromb Vasc Biol 21:1712-1719

Part IV
Poster Sessions

Part IV
Poster Sessions

Onco-suppressor p53 protein prevents an Alzheimer disease mouse model, *Pin1*-null mouse from the increase of presenilin-1.

Katsuhiko Takahashi[1,2,3], Kiyoe Shimazaki[2], Takashi Obama[1], Rina Kato[1], Hiroyuki Itabe[1] Hirotada Akiyama[3,4], Chiyoko Uchida[3,5] and Takafumi Uchida[3,4]

[1]Department of Biological Chemistry, School of Pharmaceutical Science, Showa University
[2]Department of Physiological Chemistry, School of Pharmaceutical Science, Showa University, 1-5-8, Hatanodai, Shinagawa-ku, Tokyo 142-8555, Japan
[3]Center for Interdisciplinary Research, Tohoku University, Aramaki, 6-3, Aoba, Sendai, Miyagi 980-8578, Japan
[4]Enzymology, Molecular Cell Biology, Graduate School of Agricultural Science. Tohoku University, 1-1 Amamiya, Tsutsumidori, Aoba, Sendai, Miyagi 981-8555, Japan
[5]University Health Center, Ibaraki University, 2-1 Bunkyo, Mito, Ibaraki 310-8512, Japan

Summary. The onco-suppressor p53 protein is essential for checkpoint control in response to a variety of genotoxic stresses. The DNA damage leads to phosphorylation on Ser/Thr-Pro motifs of p53, which facilitates the interaction with Pin1, a pSer/pThr-Pro-specific peptidyl prolyl isomerase. Pin1 is required for the timely accumulation and activation of p53 resulting in apoptosis or cell cycle arrest. Recently it has been indicated that *Pin1*-null (*Pin1*-/-) mouse is useful as an Alzheimer disease mouse model. To investigate the pathological relationship between Pin1 and p53, we created *Pin1*-/-*p53*-/- mice. The thymocytes in 12-week-old *Pin1*-/-*p53*-/- mice had significantly higher levels of the intracellular form of Notch1 (NIC) than the thymocytes of *p53*-/- or wild-type mice. Presenilin-1, a cleavage enzyme for NIC generation from full length Notch1 was increased in the thymocytes of *Pin1*-/-*p53*-/- mice. Presenilin-1 is counted as the enzyme for amyloid β peptides generation in the brain. In the brain of 12-week-old *Pin1*-/-*p53*-/- mice, the levels of presenilin-1 were also more than *Pin1*-/- mice, and disruption of p53 increased presenilin-1 levels of *Pin1*-/- mice. From these results, it was suggested that p53 might down-regulate presenilin-1 expression and suppress the Aβ generation from Aβ peptide precursor in *Pin1*-/- mice.

Key words. Alzheimer's disease, Notch1, p53, Pin1, Presenilin-1

The peptidyl-prolyl isomerase (PPIase), Pin1 regulates factors involved in the control of cell growth and apoptosis (Lu *et al*, 1996). Pin1 specifically associates with phosphorylated serine or threonine residue followed by proline (pS/pT-P) with WW domain, and catalyzes the *cis-trans* isomerization of the peptide bond. Isomerization modulates protein folding, biological activity, stability, and subcellular localization of its substrates (Shaw, 2002). Pin1 has been linked to cell cycle control and tumorigenesis. We created Pin1-deficient mice (Fujimori et al., 1999), and their phenotypes of Pin1-deficient mice provided us that Pin1 might implicate the pathology of cancer and Alzheimer's disease (Liou et al., 2003; Akiyama et al., 2005; Pastrino et al., 2006). Aged Pin1-deficient mice displayed tauropathy and neurodegeneration, and reduced Pin1 levels increased risk for late-onset Alzheimer's disease.

The p53 tumor suppressor gene plays a complex and critical role in maintaining genome integrity (Toledo and Wahl, 2006). The level of active p53 protein is rapidly elevated through acetylation and phosphorylation in cells with damaged DNA by various stresses including UV irradiation, DNA damaging chemicals, hypoxia, and activated oncogenes. High levels of activated p53 then drive transcription of a variety of genes that mediate the protein's biological functions for the decision to arrest cells to allow for DNA repair or eliminate the cell by apoptosis. The p53 null (p53-/-) mice were found to succumb rapidly to tumors after birth, predominantly in the thymic lymphoid lineages (Donehower *et al*, 1992).

Recently, it was reported that Pin1 is involved in the accumulation and activation of p53 in the cells with damaged DNA (Zheng *et al*, 2002; Zacchi *et al*, 2002). These reports indicated that p53 is one of the substrates of Pin1, and the effects on p53 are due to the isomeration by Pin1 (**Fig.1**). Therefore Pin1 may play a critical role in integration of the checkpoint induced by the DNA damage. These facts led us to study the physiological relation between Pin1 and p53. In order to identify the physiological relation between Pin1 and p53, we have created *Pin1-/-p53-/-* mice and found that Pin1 deletion accelerated the thymic hyperplasia in the *p53-/-* mice (Takahashi et al., 2006). The intracellular

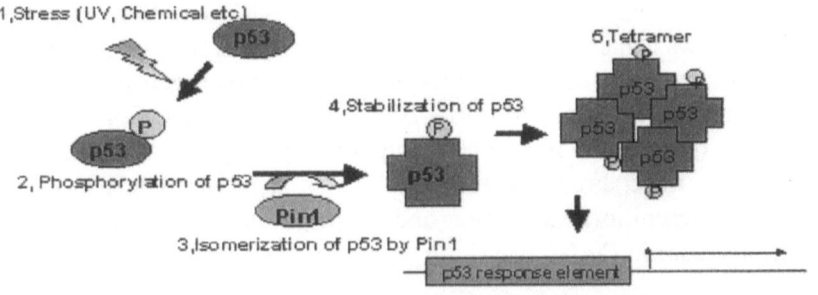

Fig.1. The relationship between Pin1 and p53 (Zheng *et al*, 2002; Zacchi *et al*, 2002).

domain of Notch1 (NIC) was significantly increased in the thymocytes of *Pin1-/-p53-/-* mice of 12 weeks age and the thymocytes may cause hyperplasia by the up-regulated NIC. Presenilin-1 level was also increased concomitantly with NIC level in the thymocytes of *Pin1-/-p53-/-* mice. And we have revealed that Pin1 could induce the degradation of NIC. These data indicate that both Pin1 and p53 are key regulators of thymocyte proliferation through Notch1 activation.

Notch1 is a transmembrane receptor that is crucial for T cell development (Radtke *et al*, 2004.). Its activated form, NIC translocates from plasma membrane to the nucleus and induces transcription of downstream genes, such as *cyclinD1*. It has been reported that p53 acts the down-regulation of presenilin1 which is a component of the γ-secretase activity necessary for cleavage of Notch1 for its activation, and p53 may negatively regulate Notch1 activation (Pastorcic and Das, 2000). Both p53 and Notch1 are linked to lymphoid malignancies and Notch1 can play a role in lymphoid oncogenesis in mice (Oswald *et al*, 2001).

Presenilin-1 is the component of γ-secretase, which can act as the cleavage protease not only for NIC from Notch-1, but also for β-amyloid (Aβ) from amyloid precursor protein (APP) (Selkoe 2001). Actually, p53 deficient mice displayed the accumulation of NIC in thymus and β-amyloid in brain (Laws and Osborne 2004; Amson et al., 2000). We recognized that presenilin-1 transcripts in Pin1-/-p53-/- thymi were more than in the other genotypes. Recently several groups indicated APP is one of the substrates of Pin1 (Akiyama et al., 2005; Pastrino et al., 2006). So, from these results, it was suggested that p53 might down-regulate presenilin-1 expression and suppress the Aβ generation from Aβ peptide precursor in *Pin1-/-* mice (**Fig. 2**).

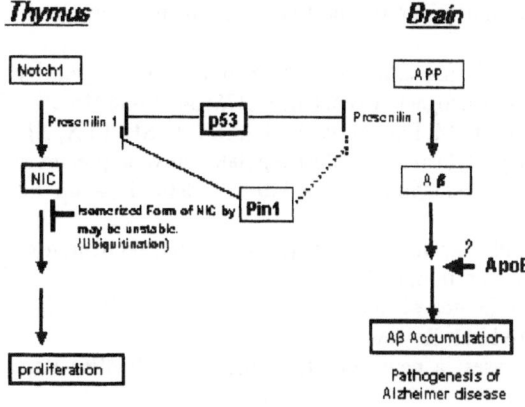

Fig.2. The scheme of Pin1-p53 network

Acknowledgement
The study in Tohoku University (T.U.) has been supported by Grant-in-Aid for Scientific Research from the Ministry of Education, Culture, Sports, Science and Technology and Grant-in-Aid for Scientific Research on Priority Areas of Japan, and in Showa University (K.T) has been supported by a Showa University Grant-in-Aid for Development of Characteristic Education from the Japanese Ministry of Education, Culture, Sports, Science and Technology.

138

References

Akiyama H, Shin RW, Uchida C, Kitamoto T, Uchida T. (2005) Pin1 promotes production of Alzheimer's amyloid beta from beta-cleaved amyloid precursor protein. Biochem Biophys Res Commun. 336: 521-529.

Amson R, Lassalle JM, Halley H, Prieur S, Lethrosne F, Roperch JP, Israeli D, Gendron MC, Duyckaerts C, Checler F, Dausset J, Cohen D, Oren M, Telerman A. (2000) Behavioral alterations associated with apoptosis and down-regulation of presenilin 1 in the brains of p53-deficient mice. Proc Natl Acad Sci U S A. 97: 5346-5350.

Donehower LA, Harvey M, Slagle BL, McArthur MJ, Montgomery CA Jr, Butel JS, Bradley A. (1992) Mice deficient for p53 are developmentally normal but susceptible to spontaneous tumours. Nature. 356: 215-221.

Fujimori F, Takahashi K, Uchida C, and Uchida T (1999) Mice lacking Pin1 develop normally, but are defective in entering cell cycle from G(0) arrest. Bioche Biophy Res Commun 265: 658-663

Laws AM, Osborne BA. (2004) p53 regulates thymic Notch1 activation. Eur J Immunol. 34: 726-734.

Liou YC, Sun A, Ryo A, Zhou XZ, Yu ZX, Huang HK, Uchida T, Bronson R, Bing G, Li X, Hunter T, and Lu KP. (2003) Role of the prolyl isomerase Pin1 in protecting against age-dependent neurodegeneration. Nature. 424: 556-561.

Lu KP, Hanes SD, Hunter T. (1996) A human peptidyl-prolyl isomerase essential for regulation of mitosis. Nature 380: 544-547.

Oswald, F., Tauber, B., Dobner, T., Bourteele, S., Kostezka, U., Adler, G., Liptay, S. and Schmid, R.M. (2001). p300 acts as a transcriptional coactivator for mammalian Notch-1 Mol. Cell. Biol. 21: 7761-7774.

Pastorcic M, Das HK. (2000) Regulation of transcription of the human presenilin-1 gene by ets transcription factors and the p53 protooncogene. J Biol Chem. 275: 34938-34945.

Pastorino L, Sun A, Lu PJ, Zhou XZ, Balastik M, Finn G, Wulf G, Lim J, Li SH, Li X, Xia W, Nicholson LK, Lu KP. (2006) The prolyl isomerase Pin1 regulates amyloid precursor protein processing and amyloid-beta production. Nature.440: 528-534. Erratum in: *Nature* 2007 446: 342

Radtke, F., Wilson, A., Mancini, S.J. and MacDonald, H.R. (2004). Notch regulation of lymphocyte development and function. Nat. Immunol. 5: 247-253.

Selkoe DJ. (2001) Presenilin, Notch, and the genesis and treatment of Alzheimer's disease. Proc Natl Acad Sci U S A. 98: 11039-11041. Review.

Shaw PE. (2002) Peptidyl-prolyl isomerases: a new twist to transcription. EMBO Rep. 3:521-526. Review.

Takahashi K, Akiyama H, Shimazaki K, Uchida C, Akiyama-Okunuki H, Tomita M, Fukumoto M, Uchida T. (2006) Ablation of a peptidyl prolyl isomerase Pin1 from p53-null mice accelerated thymic hyperplasia by increasing the level of the intracellular form of Notch1. Oncogene. In press.

Toledo F, Wahl GM. (2006) Regulating the p53 pathway: in vitro hypotheses, in vivo veritas. Nat Rev Cancer. 6: 909-923. Review.

Zacchi, P., Gostissa, M., Uchida, T., Salvagno, C., Avolio, F., Volinia, S., Ronai, Z., Blandino, G., Schneider, C. and Del Sal, G. (2002). The prolyl isomerase Pin1 reveals a mechanism to control p53 functions after genotoxic insults. Nature 419: 853-857.

Zheng, H., You, H., Zhou, X.Z., Murray, S.A., Uchida, T., Wulf, G., Gu, L., Tang, X., Lu, K.P. and Xiao, Z.X. (2002). The prolyl isomerase Pin1 is a regulator of p53 in genotoxic response. Nature 419: 849-853.

Quantification of mouse oxidized low-density lipoprotein by sandwich ELISA

Rina Kato[1], Chihiro Mori[1], Keiko Kitasato[1], Katsuhiko Takahashi[1], Satoru Arata[2], Takashi Obama[1], Hiroyuki Itabe[1]

[1]Department of Biological Chemistry, School of Pharmaceutical Science, Showa University, [2]Laboratory of DNA Recombination, Showa University, 1-5-8 Hatanodai, Shinagawa-ku, Tokyo 142-8555, Japan
<e-mail> rinakato@pharm.showa-u.ac.jp

Summary. Oxidized low-density lipoprotein (OxLDL) is one of the major factors involved in development of atherosclerosis. A close correlation exists between OxLDL levels in human circulating plasma and occurrence of cardiovascular diseases. To see insight into the patho-physiological feature of diseases, gene engineered animal models, such as apolipoprotein-E knockout mouse, are very powerful tool. We developed an ELISA procedure to measure OxLDL present in mouse circulating plasma, by modification of the previously established method. This method would be a useful tool to study the behavior and function of OxLDL in early development of atherosclerosis.

Key words. Atherosclerosis, oxidized LDL, ELISA, apoE-KO mouse, oxidized phosphatidylcholine.

Introduction

Oxidized low-density lipoprotein (OxLDL) is an important risk marker for cardiovascular diseases. OxLDL, a ligand of macrophage scavenger receptors, is very likely to induce of foam cell formation. In addition, OxLDL induces various biological responses in vascular cells such as activation of endothelial cells to promote monocyte recruitment. However, there are still several unknowns on the behavior of OxLDL in vivo.

Measurement of OxLDL present in human circulation has been achieved by our sandwich ELISA procedure using a monoclonal antibody (mAb) recognizing oxidized phosphatidylcholine (OxPC), DLH3, and anti-human apolipoprotein B (apoB) polyclonal antibody (pAb) (Itabe et al. 1994, Itabe et al. 1996, Ehara et al. 2001). In addition to ours, several other groups have raised anti-OxLDL mAbs and reported ELISA methods to measure OxLDL present in human circulation (summarized in a review; Itabe and Ueda 2007). A number of clinical studies using these ELISA procedures demonstrated that increased OxLDL levels were observed in plasma from patients with cardiovascular diseases, and a close correlation between plasma OxLDL levels and severity of the diseases was strongly suggested.

Apolipoprotein E-knockout (apoE-KO) mouse is a good model animal for atherosclerosis and is now used in worldwide (Zhang et al. 1992). Lack of apoE causes severe hypercholesterolemia and spontaneous development of atherosclerosis in aorta. In addition to clinical and epidemiological studies to analyze the plasma OxLDL in patients, genetic approach utilizing gene engineered animal models such as apoE-KO mouse should be useful to explore the mechanisms of atherogensis and precise role of OxLDL in the diseases. Here, we report a procedure to measure OxLDL in mouse circulating plasma.

Results and Discussion

We previously developed a sandwich ELISA procedure to determine OxLDL in human plasma using a mAb DLH3 and an anti-human apoB pAb (Itabe et al. 1996). Since DLH3 recognizes OxPC generated in OxLDL, it reacts with mouse OxLDL as well as human OxLDL, but the anti-human apoB pAb used in the assay has species specificity. To solve this problem, an antibody against mouse apoB pAb was needed. We immunized rabbits with SDS-PAGE-purified mouse apoB-48, a short version of apoB protein, which is the major component in murine LDL. This anti-mouse apoB-48 pAb binds equally well with mouse native LDL and OxLDL.

The ELISA procedure for measuring mouse OxLDL was basically achieved by replacing anti-apoB antibody (Fig. 1). Briefly, 96 well-microtiter wells were coated with 100 μl of mAb DLH3 (5 μg/ml) at room temperature for 2 hr. Then the wells were blocked with BSA using Tris-buffered saline (TBS) containing 2% BSA. Mouse LDL fractions (100 μl/well; 100 μg/ml) were added to the wells and incubated for one night. The wells were washed with TBS containing 0.01% Tween-20 for 3 times. The wells were incubated with rabbit anti-mouse apoB-48 pAb (1/2,000 dilution with TBS containing 2% skimmed milk) at room temperature followed by incubation with donkey ALP-conjugated anti-rabbit IgG pAb

(1/4,000 dilution) at 37°C for 1 hr. To visualize the reaction, 100 μl of p-nitrophenylphosphate solution (1 mg/ml) were added to the wells and incubated at the 37°C for 10-30 min. Absorbance at 405 nm was measured using a microtiter plate.

In every microtiter plate several different amounts of copper-induced mouse OxLDL (0.01-5 ng/well) were measured together with the samples to obtain a standard curve. LDL was separated from plasma of apoE-knockout mice by sequential ultracentrifugation. OxLDL was prepared by incubating the mouse LDL with 5 μM $CuSO_4$ at 37°C for 3hr and used as standard. Since the volume of plasma available from a mouse is limited, we maximized the experimental conditions to improve the sensitivity of the assay as much as 5-fold.

Using the mouse version of the ELSIA procedure, 0.05 ng of mouse OxLDL in a well can be detected. We determined the plasma OxLDL levels in male apoE-KO mice (10 weeks old) was 0.015 ± 0.004 ng OxLDL / μg LDL protein. It is interesting that OxLDL level in apoE-KO mice is about 10-times lower than normal human OxLDL level, which is about 0.10 ng/mg LDL protein (Iateb and Ueda 2007). Of cause it is difficult to compare plasma OxLDL levels in 10-week-old mice and about

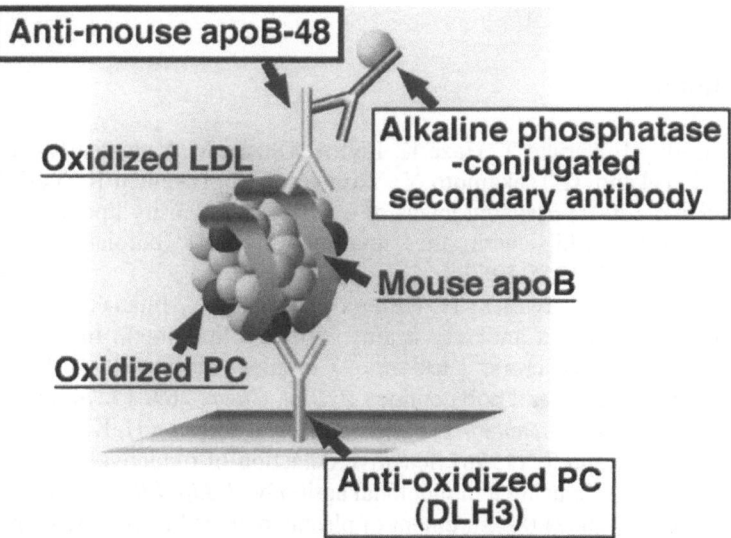

Fig. 1 Schematic Illustration of the sandwich ELISA system for measuring mouse plasma OxLDL. Anti-OxLDL mAb DLH3 was precoated onto microtiter wells. OxLDL particles are captured by DLH3 mAb, and then the amount of OxLDL was measured using anti-mouse apoB-48 rabbit pAb and enzyme-conjugated anti-rabbit IgG antibody.

50 year-old men directly. There are big difference in lipoprotein metabolism between mouse and human. Moreover, apoE-KO mouse is a hypercholesterolemic, and the human subjects of the OxLDL measurement are normal people. Even though, it is possible to think that plasma OxLDL levels might be affected by various environmental stresses, because the mice maintained in an SPF facility should be free from temperature change and bacterial infections.

The plasma OxLDL levels in apoE-KO mice increased to 0.048 $\pm$ 0.010 ng/mg LDL at the week 20, which is 3 times higher than that of 10-week old mice. This increase in OxLDL seems to be a specific change since the OxLDL levels at 28 and 40 weeks of age were as low as the level of the week 10. The mice were maintained on a chow diet throughout the experiments. Plasma triglyceride concentration of the mice was not changed significantly, and total cholesterol concentration was highest at the week 28 that was1.5 times higher than that of the week 10.

We would like to conclude the following three points: First, measurement of mouse plasma OxLDL levels is technically achieved. Second, the basal levels of mouse plasma OxLDL seems to be very low comparing with normal human subjects. Third, plasma OxLDL levels in apoE-KO mouse can be markedly changed during the aging, suggesting that OxLDL can be induced under certain physiological conditions.

References

Ehara S, Ueda M, Naruko T, Haze K, Itoh A, Otuska M, Komatsu R, Matsuo T, Itabe H, Takano T, Tsukamoto Y, Yoshiyama M, Takeuchi K, Yoshikawa J, Becker AE. (2001) Elevated levels of oxidized low density lipoprotein show a positive relationship with the severity of acute coronary syndromes. *Circulation.* **103:** 1955-1960.

Itabe H, Takeshima E, Iwasaki H, Kimura J, Yoshida Y, Imanaka T, Takano T. (1994) A monoclonal antibody against oxidized lipoprotein recognizes foam cells in atherosclerotic lesions. Complex formation of oxidized phosphatidylcholine and polypeptides. *J. Biol. Chem.* **269:** 15274-15279.

Itabe H, Yamamoto H, Imanaka T, Shimamura K, Uchiyama H, Kimura J, Sanaka T, Hata Y, Takano T. (1996) Sensitive detection of oxidatively modified low density lipoprotein using a monoclonal antibody. *J. Lipid Res.* **37:** 45-53.

Itabe H, Ueda M. (2007) Measurement of plasma oxidized low-density lipoprotein and its clinical implications. *J. Atheroscler. Thromb.* **14:** 1-11.

Zhang SH, Reddick RL, Piedrahita JA, Maeda N. (1992) Spontaneous hypercholesterolemia and arterial lesions in mice lacking apolipoprotein E. Science. 258: 468-471.

Regulation of sPLA$_2$-IIA expression in cytokine-stimulated rat fibroblasts

Hiroshi Kuwata, and Ichiro Kudo

Department of Health Chemistry, School of Pharmaceutical Sciences, Showa University, 1-5-8 Hatanodai, Shinagawa-ku, Tokyo 142-8555, Japan
<e-mail>ichi-ku@pharm.showa-u.ac.jp

Summary. Phospholipase A$_2$ (PLA$_2$) enzymes are family of enzymes, which catalyze hydrolysis of glycerophospholipids at the *sn*-2 position, generating free fatty acids, including arachidonic acid (AA), and lysophospholipids. The AA is converted into various bioactive lipid mediators, such as prostaglandins and leukotrienes, through the cyclooxygenase or 5-lipoxygenase (LOX) pathway. Although it has been thought that group IVA cytosolic PLA$_2\alpha$, which can preferentially hydrolyze AA containing glycerophospholipids, plays a role in the stimulus-dependent lipid mediator biosynthesis, the roles of other PLA$_2$ enzymes are unclear.

Key words. sPLA$_2$-IIA, cPLA$_2\alpha$, iPLA$_2\gamma$, 12/15-lipoxygenase, cPLA$_2$ inhibitor

1 Introduction

Phospholipase A_2s (PLA_2) represent the first step of arachidonate metabolism, regulating the generation of various bioactive lipid mediators, including prostaglandins and leukotrienes. Currently, more than 20 mammalian PLA_2 enzymes have been identified and classified into several families, including secretory PLA_2 ($sPLA_2$), cytosolic PLA_2 ($cPLA_2$), and calcium-independent PLA_2 ($iPLA_2$). Among them, group IIA $sPLA_2$ ($sPLA_2$-IIA) is an inducible enzyme, which is found in the circulation and in the tissues in associated with a variety of pathological conditions such as rheumatoid arthritis, sepsis, and atherosclerosis. Several previous studies have shown that the expression of $sPLA_2$-IIA induced by proinflammatory stimuli is attenuated by group IVA $cPLA_2$ ($cPLA_2\alpha$) or lipoxygenase (LOX) inhibitors, leading the proposal that the lipid product(s) produced via the PLA_2-LOX pathway might be involved in the stimulus-dependent $sPLA_2$-IIA induction (Kuwata et al. 1998) (Couturier et al. 1999) (Ni et al. 2006). However, the mechanisms underlying the regulation of the $sPLA_2$-IIA expression are not fully understood. In order to understand such the mechanisms, we studied the regulation of $sPLA_2$-IIA expression in cytokine-stimulated rat fibroblastic 3Y1 cells.

2 Group IVA $cPLA_2\alpha$ is not involved in the regulation of the $sPLA_2$-IIA expression in cytokine-stimulated 3Y1 cells

Activation of rat fibroblastic 3Y1 cells with the proinflammatory cytokines, interleukin-1β (IL-1β) and tumor necrosis factor α (TNFα), elicited the *de novo* induction of the $sPLA_2$-IIA. The inducible expression of $sPLA_2$-IIA mRNA was markedly attenuated by the addition of the $cPLA_2$ inhibitor arachidonyl trifluoromethyl ketone (AACOCF$_3$), implying that AACOCF$_3$-sensitive PLA_2, such as $cPLA_2\alpha$, might be involved in the cytokine-dependent $sPLA_2$-IIA mRNA expression (Kuwata et al. 1998). To confirm this possibility, we established $cPLA_2\alpha$-knockdown 3Y1 cells, in which expression of $cPLA_2\alpha$ protein was undetectable. Unexpectedly, the cytokine-dependent induction of $sPLA_2$-IIA in $cPLA_2\alpha$-knockdown cells was comparable to that replicate control cells, conflicting with pharmacological study of $cPLA_2\alpha$. Furthermore, the inducible expression of $sPLA_2$-IIA in $cPLA_2\alpha$-knockdown cells was markedly attenuated by AACOCF$_3$. These results suggest that $cPLA_2\alpha$ is not involved in the regulation of the $sPLA_2$-IIA expression under these conditions (Kuwata et al. 2007).

3 Group VIB iPLA$_2\gamma$ participates in the regulation of the sPLA$_2$-IIA expression in 3Y1 cells

Because AACOCF$_3$ is now known to inhibit several enzymes including iPLA$_2$s, we speculated that enzymes in the iPLA$_2$ family could be involved in the regulation of the cytokine-dependent sPLA$_2$-IIA expression. To assess such possibility, we examined the effects of iPLA$_2$ inhibitor bromoenol lactone (BEL) on the cytokine-induced sPLA$_2$-IIA expression. The inducible expression of sPLA$_2$-IIA was markedly attenuated by BEL, implying that iPLA$_2$ enzymes may participate in the regulation of the sPLA$_2$-IIA expression. We next examined the existence of the iPLA$_2$ enzymes in 3Y1 cells. RNA blotting revealed that mRNA expressions of iPLA$_2\beta$ and iPLA$_2\gamma$ were evident, whereas the expressions of other iPLA$_2$ enzymes were undetected, suggesting that iPLA$_2\beta$ and iPLA$_2\gamma$ are the dominant iPLA$_2$ enzymes expressed in 3Y1 cells. To identify which iPLA$_2$ enzyme is involved in the sPLA$_2$-IIA expression, we used the selective inhibitor of iPLA$_2\beta$ and iPLA$_2\gamma$. We found that treatment of the iPLA$_2\gamma$ inhibitor (R)-BEL, but not the iPLA$_2\beta$ inhibitor (S)-BEL, markedly attenuated the sPLA$_2$-IIA expression, implying that iPLA$_2\gamma$ might be involved in the regulation of sPLA$_2$-IIA expression. In support with this possibility, the enzymatic activity of iPLA$_2\gamma$ was potently inhibited by AACOCF$_3$ in a dose-dependent manner. To confirm this possibility, we identify iPLA$_2\gamma$-knockdown 3Y1 cells. We found that knockdown of iPLA$_2\gamma$ markedly reduced the cytokine-dependent sPLA$_2$-IIA expression. These results indicated that iPLA$_2\gamma$ participated in the sPLA$_2$-IIA expression in 3Y1 cells. (Kuwata et al. 2007).

4 Involvement of 12/15-LOX in the regulation of the sPLA$_2$-IIA expression

In order to know which LOX enzyme is participated in the cytokine-dependent sPLA$_2$-IIA expression, RT-PCR was conducted using specific primers for 5-LOX and 12/15-LOX. RT-PCR analysis revealed that 12/15-LOX, but not 5-LOX, were present in 3Y1 cells. Furthermore, stimulation of 12/15-overexpressing 3Y1 cells led to marked increase in the sPLA$_2$-IIA expression, compared to replicate control cells. These results suggest that 12/15-LOX is also involved in the regulation of the cytokine-dependent sPLA$_2$-IIA expression (Kuwata et al. 2000).

146

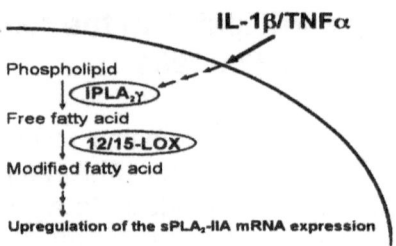

Fig. 1. Schematic illustration of the signaling pathway for the sPLA$_2$-IIA expression in cytokine-stimulated 3Y1 cells.

Conclusion

On the basis of our results, we are considering the following model (Fig.1). After stimulation of 3Y1 cells with IL-1β and TNFα, certain lipid product(s) was produced via the iPLA$_2$γ-12/15-LOX pathway. The lipid product(s), in turn, activates the signaling pathway, which leads to induce the sPLA$_2$-IIA expression.

References

Couturier C, Brouillet A, Couriaud C, Koumanov K, Bereziat G, Andreani M (1999) Interleukin 1β induces type II-secreted phospholipase A$_2$ gene in vascular smooth muscle cells by a nuclear factor κB and peroxisome proliferator-activated receptor-mediated process. J Biol Chem 274 23085-23093

Kuwata H, Nakatani Y, Murakami M, Kudo I (1998) Cytosolic phospholipase A$_2$ is required for cytokine-induced expression of type IIA secretory phospholipase A$_2$ that mediates optimal cyclooxygenase-2-dependent delayed prostaglandin E$_2$ generation in rat 3Y1 fibroblasts. J Biol Chem 273 1733-1740

Kuwata H, Fujimoto C, Yoda E, Shimbara S, Nakatani Y, Hara S, Murakami M, Kudo I (2007) A novel role of group VIB calcium-independent phospholipase A$_2$ (iPLA$_2$γ) in the inducible expression of group IIA secretory phospholipase A$_2$ in rat fibroblastic cells. J Biol Chem in press.

Kuwata H, Yamamoto S, Miyazaki Y, Shimbara S, Nakatani Y, Suzuki H, Ueda N, Yamamoto S, Murakami M, Kudo I (2000) Studies on a mechanism by which cytosolic phospholipase A$_2$ regulates the expression and function of type IIA secretory phospholipase A$_2$. J Immunol 165 4024-4031

Ni Z, Okeley NM, Smart BP, Gelb MH (2006) Intracellular actions of group IIA secreted phospholipase A$_2$ and group IVA cytosolic phospholipase A$_2$ contribute to arachidonic acid release and prostaglandin production in rat gastric mucosal cells and transfected human embryonic kidney cells. J Biol Chem 281 16245-16255

Comparison Between Small Dense LDL-Cholesterol and LDL-Cholesterol to Predict Coronary Events in Stable Coronary Heart Disease

Yuuya Yokota[1], Shinji Koba[1], Fumiyoshi Tsunoda[1], Yoshihisa Ban[1], Takayuki Sato[1], Makoto Shoji[1], Hiroshi Suzuki[1], Takashi Katagiri[1]

[1]Third Department of Internal Medicine, Showa University School of Medicine, 1-5-8 Hatanodai, Shinagawa-ku, Tokyo 142-8666, Japan

Summary. Recent evidence suggests that small dense low-density lipoprotein (sd-LDL) particles are more atherogenic than large-LDLs in spite of their lower cholesterol contents. This study aimed to determine whether sd-LDL-cholesterol (sd-LDL-C) is superior to LDL-C as a biomarker of the severe coronary heart disease (CHD). We compared the LDL particle size by gradient gel electrophoresis and sd-LDL-C concentrations quantified by heparin-magnesium precipitation in two groups: 482 consecutive patients with stable CHD who had undergone coronary arteriography and 389 non-diabetic subjects without CHD who were not receiving any lipid-lowering drugs. The LDL size, large-LDL-C (estimated by subtracting the sd-LDL-C concentration from the LDL-C concentration), and HDL-C were significantly lower in the CHD subjects than in the healthy subjects, and the sd-LDL-C was significantly higher, in both men and women. The LDL-C was modestly higher and the sd-LDL-C was significantly higher in 258 patients with coronary events (defined as coronary revascularization therapy) than in the patients without events, irrespective of treatment by LDL-lowering drugs. Large-LDL-C, in contrast, was similar between the two groups. Multivariate logistic regression analysis revealed that sd-LDL-C levels were significantly associated with coronary events independently of LDL-C, HDL-C, and high-sensitivity CRP. The sd-LDL-C levels are more powerful than LDL-C levels as disease markers for the determination of high-risk patients among patients with stable CHD.

Key words. Small dense LDL, coronary heart disease, large LDL, LDL-cholesterol

1 Introduction

Small dense (sd)-LDL particles have been suggested to be highly athero-genic compare to normal LDLs (Berneis et al. 2002). A recent report from the Québec cardiovascular study has confirmed that a greater proportion of sd-LDL at baseline is a strong and independent predictor of CHD in the first 7 years of follow-up (St-Pierre et al. 2005). In contrast, an elevated concentration of large-LDL is a poor predictor of CHD and seems to be associated with a low CHD risk. This suggests that the atherogenicity of LDL differs among heterogeneous LDL particles and that the association of atherogenic LDL-C with CHD is chiefly due to the sd-LDL component.

Hirano and co-workers have recently discovered a simple and rapid method for measuring sd-LDL-C by heparin magnesium precipitation (Hirano et al. 2005). The aims of this study were to determine whether sd-LDL-C is superior to LDL-C as a biomarker of CHD progression among healthy non-CHD subjects and stable CHD subjects with and without lipid-lowering drugs.

2 Methods

2.1 Subjects

The study enrolled 482 consecutive previously diagnosed CHD patients who underwent coronary arteriography at our hospital from April 2004 to September 2005 and 389 non-CHD healthy subjects who did not receive lipid-lowering drugs. The CHD group consisted of patients with previous MI (226 men and 47 women) and/or stable angina pectoris (216 men and 55 women). Two hundred and eighty male patients and 58 female patients had previously undergone the coronary revascularization procedure. The diagnosis of hypertension was based on a history of hypertension or blood pressure above 140 mmHg systolic or 90 mmHg diastolic. Diabetes mellitus was diagnosed as a fasting serum glucose value greater than 126 mg/dL, hemoglobin (Hb) A1c levels greater than 6.5%, or treatment with either oral hypoglycemic agents or insulin. The institutional review board of Showa University approved this protocol. The investigation conformed to the principles of the Declaration of Helsinki.

2.2 Lipoprotein and Inflammatory Parameter Analysis

The fasting blood samples were obtained by venipuncture and the serum was stored at 4°C and used for the assay within 3 days after sampling. LDL-C was measured by direct homogenous assay of the serum using detergents (LDL-EX, Denka Seiken, Tokyo, Japan). The peak LDL particle diameter was determined by 2-16% non-denatured polyacrylamide gel electrophoresis according to the method of Nichols, et al. (Nichols et al. 1986). The sd-LDL phenotype was defined as a diameter equal to or less than 255 Å (Berneis et al. 2002). The sd-LDL-C was measured by rapid assay using a modified version of the heparin-magnesium precipitation method (Hirano et al. 2005). The high sensitive CRP was measured by the Dade Behring BN assay. Large-LDL-C was estimated by subtracting the sd-LDL-C concentration from the LDL-C concentration. Non-HDL-C was estimated by subtracting the HDL-C concentration from the total-cholesterol concentration.

2.3 Definition of Coronary Events and the Severity of Coronary Atherosclerosis

Coronary events were defined as coronary revascualrization with either percutaneous coronary intervention or coronary-artery bypass grafting. The severity of coronary atherosclerosis was estimated by calculating the Gensini score, an established method for grading CHD severity (Gensini 1983).

2.4 Statistics

Statistical analyses were completed using Statview 5.0 (SAS Institute, Cavy, NC). Comparisons among the healthy groups and subgroups of CHD based on the presence of lipid-lowering drug were performed by one way ANOVA. The Bonferroni/Dunn post-hoc test was used when a significant group effect was observed. Comparisons of mean values between the two groups were performed by the Student's t-test. Logistic regression analysis was used to assess the independent association between serum markers and the incidence of coronary events.

3. Results

Table 1 compares the general characteristics and serum lipid markers between the healthy subjects and CHD patients with and without lipid-lowering drugs (data for both sexes shown). LDL-C and large-LDL-C were markedly lower in the CHD patients with lipid-lowering therapy than in the un-medicated CHD patients and the healthy subjects of both sexes. Sd-LDL-C levels were significantly higher and large-LDL-C levels were significantly lower in the CHD patients without lipid-lowering drugs than in the healthy subjects of both sexes, whereas the LDL-C levels were similar between the two groups. The LDL particle size and HDL-C levels were significantly lower in the CHD patients than in the healthy subjects of both sexes, irrespective of the presence of lipid-lowering drugs.

Table 1 Comparison of lipid parameters between control and CHD subjects

	Men				Women			
	Control	CHD Lipid-lowering		p	Control	CHD Lipid-lowering		p
		(-)	(+)			(-)	(+)	
Number	226	180	206		163	36	60	
Age	60.5 ± 12.7	67.9 ± 10.4*	63.2 ± 11.2*†	<0.0001	57.6 ± 11.5	68.8 ± 9.8*	70.3 ± 6.8*	<0.0001
BMI	23.6± 3.3	23.4 ± 3.2	24.9 ± 3.2*†	<0.0001	22.9 ± 3.9	23.4 ± 4.8	23.4 ± 3.5	NS
Hypertension	54%	77%	80%	<0.0001	63%	89%	80%	0.0014
Diabetes	0	42%	39%	<0.0001	0	39%	43%	<0.0001
LDL size, Å	257.3± 3.7	256.4 ± 3.9*	256.3 ± 4.7*†	<0.0001	258.6 ± 4.6	256.4 ± 3.8*	256.4 ± 3.8*	0.0005
LDL-C	126.6 ± 32.9	121.1 ± 33.6	112.0 ± 26.2*†	<0.0001	133.9± 37.2	127.7 ± 36.8	105.5 ± 24.0*†	<0.0001
Sd-LDL-C	28.9 ± 15.9	36.6 ± 21.6*	33.3 ± 17.4*	<0.0001	26.6 ± 16.2	37.0 ± 21.6*	31.2 ± 18.2	0.0033
Large-LDL-C	97.7 ± 27.4	84.3 ± 27.2*	78.7 ± 23.3*	<0.0001	106.9± 31.6	90.0 ± 30.0*	74.3 ± 18.7*†	<0.0001
HDL-C	58.5 ± 14.5	44.5 ± 12.6*	44.2 ± 11.6*	<0.0001	65.6 ± 18.6	49.3 ± 15.5*	52.9 ± 14.4*	<0.0001
Triglyceride	129.8 ± 79.0	130.8 ± 77.0	139.8 ± 65.9	NS	101.8± 66.9	128.5 ± 74.2	113.0 ±69.4	NS

Table 2 shows the general characteristics and various serum parameters in patients divided into four groups based on the coronary events and the presence of lipid-lowering drugs. Patients with events exhibited significantly higher levels of LDL-C, non-HDL-C, and apo B, significantly lower levels of HDL-C and apo A-1, irrespective of the presence of lipid-lowering drugs. The LDL-C, sd-LDL-C, and large-LDL-C levels were significantly different among the four groups, and the sd-LDL-C levels were significantly higher in the patients with coronary events, irrespective of the presence of lipid-lowering drugs (Figure 1).

Table 2. Comparison of lipid parameters in CHD patients with or without Lipid-lowering drugs

	With Lipid-lowering (N=266)				Without lipid lowering (N=216)		
	Mild	p	Severe	p	Mild	p	Severe
Male/Female	102 / 31	NS	104 / 29	NS	72 / 19	NS	108 / 17
Age	64.5 ± 0.9	NS	65.1 ± 1.0	NS	67.2 ± 1.1	NS	66.6 ± 0.9
HbA1c, %	5.9 ± 0.1	NS	6.2 ± 0.1	0.0401	5.8 ± 0.1	0.0069	6.3 ± 0.1
LDL size, Å	256.1 ± 0.4	0.0447	255.0 ± 0.4	0.0338	256.3 ± 0.5	NS	256.5± 0.3
LDL-C, mg/dL	105.3 ± 2.2	0.0016	115.8 ± 2.5	NS	114.1 ± 3.2	0.0024	128.3 ± 3.?
Non-HDL-C, mg/dL	130.4 ± 2.4	0.0171	139.3 ± 2.8	NS	142.0 ± 3.5	0.0402	152.2 ± 3.?
HDL-C, mg/dL	49.3 ± 1.0	<0.0001	43.0 ± 1.1	0.0059	48.4 ± 1.7	0.0031	42.9 ± 1.0
Triglyceride, mg/dL	124.3 ± 4.8	0.0234	143.1 ± 6.7	NS	124.5 ± 8.1	NS	135.4 ± 6.?
Apo A-1, mg/dL	129.1 ± 1.9	<0.0001	116.2 ± 2.1	0.0198	124.9 ± 2.9	0.0023	114.2 ± 2.?
Apo B, mg/dL	84.7 ± 1.5	0.0019	92.2 ± 1.9	NS	86.8 ± 2.2	0.0013	99.4 ± 2.2
hs-CRP, mg/L	1.49 ± 0.13 (n=127)	NS	1.47 ± 0.14 (n=119)	NS	1.49 ± 0.15 (n=86)	0.0276	2.07 ± 0.1? (n=119))
Gensini score	18.6 ± 1.6	<0.0001	62.3 ± 3.2	<0.0001	18.7 ± 1.5	<0.0001	59.7 ± 3.6

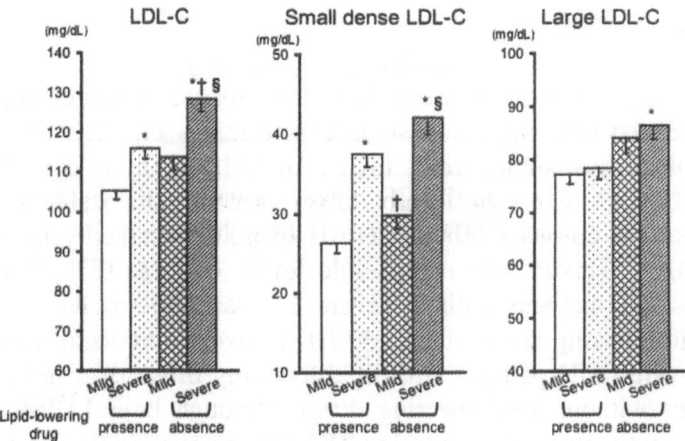

Fig. 1 Comparison of LDL-C levels, sd-LDL-C levels, and large-LDL-C levels in patients divided into four groups based on the severity of CHD and the presence of lipid-lowering drugs. LDL-C (F=12.78, p<0.0001), sd-LDL-C (F=14.52, p<0.0001) and large LDL-C (F=3.97, p=0.0082) levels were significantly different among the four groups by ANOVA. Data expresses as mean minus standard error. *p<0.0083 vs mild CHD with lipid-lowering drugs, †p<0.0083 vs severe CHD with lipid-lowering drugs, §p<0.0083 vs mild CHD without lipid-lowering drugs by Bonferroni/Dunn post-hoc test. ☐ Mild-CHD with lipid-lowering drugs, ▨ severe-CHD with lipid-lowering drugs, ▧ mild-CHD without lipid-lowering drugs, ▨ severe-CHD without lipid-lowering drugs

According to our multivariate logistic regression analysis to compare sd-LDL-C with well-established potent risk factors for determining the progression of CHD, an elevated sd-LDL-C concentration was significantly associated with coronary events independently of the levels of LDL-C, HDL-C, apo B, non-HDL-C, and high sensitive CRP in the stable CHD patients and in the CHD patients un-medicated by lipid-lowering drugs (Table 3).

Table 3. Multivariate Logistic regression analysis for determinants of severe CHD

Model 1	Overall CHD			Non-lipid-lowering group			Lipid-lowering group		
	Odds	95% CI	p	Odds	95% CI	p	Odds	95% CI	P
LDL-C	1.011	0.998-1.023	NS	1.009	0.991-1.027	NS	1.013	0.996-1.030	NS
Sd-LDL-C	1.024	1.008-1.040	0.0034	1.029	1.006-1.052	0.0144	1.018	0.996-1.040	NS
HDL-C	0.977	0.96-0.995	0.0122	0.989	0.963-1.015	NS	0.968	0.944-0.991	0.0081
Non-HDL-C	0.991	0.980-1.002	NS	0.990	0.973-1.006	NS	0.992	0.977-1.007	NS
hs-CRP	2.801	0.805-9.752	NS	8.368	0.876-38.89	NS	0.830	0.136-5.046	NS
Model2									
LDL-C	1.009	0.996-1.022	NS	1.005	0.987-1.023	NS	1.013	0.995-1.031	NS
Sd-LDL-C	1.022	1.006-1.037	0.0066	1.028	1.005-1.052	0.0189	1.016	0.995-1.031	NS
HDL-C	0.975	0.958-0.993	0.0071	0.986	0.960-1.012	NS	0.967	0.944-0.991	0.0072
Apo B	0.992	0.974-1.011	NS	0.994	0.969-1.021	NS	0.992	0.966-1.018	NS
hs-CRP	2.404	0.680-8.496	NS	5.838	0.876-38.89	NS	0.888	0.146-5.386	NS

4 Discussion

The present study using a simple method for the measurement of sd-LDL-C produced three new and significant findings on LDL subclasses: first, sd-LDL-C was significantly higher in CHD than in non-CHD whereas large-LDL-C was significantly lower; second, the stable CHD patients with severe coronary atherosclerosis exhibited markedly higher levels of sd-LDL-C, versus only comparable levels of large-LDL-C, compared to stable CHD patients without severe coronary atherosclerosis, irrespective of lipid lowering drugs; third, sd-LDL-C levels were significantly higher in the severe CHD patients with lipid-lowering drugs than in the mild CHD patients without lipid-lowering drugs, although large-LDL-C levels were higher in the unmedicated mild CHD patients than in the medicated severe CHD patients. Thus, the sd-LDL-C concentration seems to be a better surrogate marker than the LDL-C concentration for predicting the progression of CHD.

Our preliminary study was the first to demonstrate changes in sd-LDL mass in relation to increased severities determined by coronary scoring systems (Koba et al. 2006). Our present study confirmed this preliminary result in large population of stable CHD patients. The study also compared LDL-C, sd-LDL-C, non-HDL-C, and apo B to determine which were more closely associated with progressive CHD. Non-HDL-C and apo B was reported to be more predictive of CHD events than LDL-C, particularly among patients with elevated triglycerides, in the US cholesterol educational program Adult Treatment of Panel III. Our study, on the other hand, identified sd-LDL as the most powerful determinant of coronary events. The strong associations of sd-LDL-C with both LDL-C and triglyceride suggest that sd-LDL may be a better surrogate marker for severe coronary atherosclerosis than LDL-C, triglyceride, non-HDL-C, or apo B.

5 Conclusion

Sd-LDL particles and a high concentration of sd-LDL-C were both potent risk factors for CHD. When defining CHD progression by various clinical features and angiographic findings, our results indicated that a high sd-LDL-C concentration was closely related to the CHD severity independently of classical coronary risk factors, whereas a high large-LDL-C concentration was not. These suggest that the progression of CHD is closely linked not to the large LDL particles, but to the amount of sd-LDL.

References

Berneis KK, Krauss RM (2002) Metabolic origins and clinical significance of LDL heterogeneity. J Lipid Res 43:1363-1379.

Gensini GG (1983) A more meaningful scoring system for determining the severity of coronary heart disease. Am J Cardiol 51:606.

Hirano T, Ito Y, Yoshino G (2005) Measurement of small dense low-density lipoprotein particles. J Atheroscler Thromb 12:67-72.

Koba S, Hirano T, ItoY, Tsunoda, T, Yokota Y, Ban Y, Iso Y, Suzuki H, Katagiri T (2006) Significance of small dense low-density lipoprotein-cholesterol concentrations in relation to the severity of coronary heart diseases. Atherosclerosis 189:206-214.

Nichols AV, Krauss RM, Musliner TA (1986) Nondenaturing polyacrylamide gradient gel electrophoresis. Methods Enzymol 128:417-33.

St-Pierre AC, Cantin B, Dagenais GR, Mauriége P, Bernard PM, Després JP, Lamarche B (2005) Low-density lipoprotein subfractions and the long-term risk of ischemic heart disease in men, 13-year follow-up data from the Québec cardiovascular study. Arterioscler Thromb Vasc Biol 25:553-559.

Acceleration of foam cell formation by leptin in human monocyte-derived macrophages

Shigeki Hongo, Takuya Watanabe, Keiko Takahashi, and Akira Miyazaki

Department of Biochemistry, Showa University School of Medicine, 1-5-8 Hatanodai, Shinagawa-ku, Tokyo 142-8555, Japan

Summary. Leptin is an adipose tissue-derived hormone implicated in the pathogenesis of atherosclerosis. Peripheral blood monocytes were incubated for 7 days to induce differentiation into macrophages. Under these conditions, exogenous leptin was added to examine its effect on macrophage foam cell formation. Leptin (5 nmol/L) significantly increased acetylated LDL-induced cholesteryl ester accumulation. Leptin increased both acyl-coenzyme A: cholesterol acyltransferase (ACAT) activity and ACAT-1 protein expression. Among the four ACAT-1 mRNA transcripts, two shorter transcripts (2.8- and 3.6-kb) were up-regulated upon leptin treatment. Importantly, leptin receptor expression was up-regulated during monocytic differentiation into macrophages, indicating stage-specific leptin signaling in human monocyte-macrophages.

Key words. Leptin, Leptin receptor, Monocyte-macrophages, Foam cells, ACAT-1, Human

1 Introduction

Leptin is a 16-kDa peptide hormone secreted by adipose tissues that targets the hypothalamus to regulate appetite and energy expenditure (Zhang et al. 1994). However, the concentration of circulating leptin is elevated in obese human subjects, suggesting leptin resistance in obesity. Elevation of plasma leptin has been suggested to promote cardiovascular diseases, including atherosclerosis (Beltowski 2006). Accumulation of excessive amounts of cholesteryl ester in cells is a hallmark of early atherosclerotic lesions (Glass and Witztum 2001). Intracellular cholesterol esterification is catalyzed by acyl-coenzyme A: cholesterol acyltransferase (ACAT). ACAT-1 is expressed at high levels by macrophage-derived foam cells in atherosclerotic lesions and up-regulated during differentiation of human monocytes into macrophages *in vitro* (Miyazaki et al. 1998).

To examine whether leptin directly affects human monocyte-macrophages to promote atherosclerosis, we studied the effects of leptin on cholesteryl ester accumulation induced by acetyl-LDL and ACAT-1 expression in primary culture systems. Further, we first analyzed the time-dependent changes in the expression of leptin receptor (Ob-R) protein in primary cultured human monocytes during differentiation into macrophages.

2 Methods

Human peripheral mononuclear cells were isolated from the blood of healthy volunteers by Ficoll density gradient centrifugation. Purified monocytes were suspended in RPMI 1640 and seeded onto 6-cm dishes (4 $\times 10^6$ cells/dish). After 1 h of incubation for adherence, the medium was replaced with RPMI 1640 supplemented with 10% pooled human serum. Adhered monocytes were incubated at 37°C in 5% CO_2 for seven days to induce differentiation into macrophages in the presence or absence of human recombinant leptin.

Cholesterol esterification assay was conducted by determining the radioactivity of cholesteryl [^{3}H]oleate after incubating monocyte-macrophages with acetylated low density lipoprotein (acetyl-LDL) and [^{3}H]oleate conjugated with BSA.

Expressions of mRNA and protein of ACAT-1 were studied by Northern and Western blotting analyses, respectively. The ACAT activity was determined by the reconstituted assay method.

3 Results

3.1 Leptin increases acetyl-LDL-induced cholesteryl ester accumulation and ACAT-1 expression in human monocyte-derived macrophages

Acetyl-LDL-induced accumulation of cholesteryl ester in differentiated macrophages increased in a dose-dependent manner. Furthermore, leptin (5 nmol/L) significantly increased acetyl-LDL (5 and 10 μg/mL)-induced cholesteryl ester accumulation by ~1.5-fold. Leptin (5 nmol/L) had no significant effect on endocytic uptake of acetyl-LDL or ABCA1 (ATP-binding cassette transporter A1) protein expression.

Human monocytes were incubated for 7 days in the absence or presence of exogenous recombinant leptin at various concentrations. Leptin increased the expression of ACAT-1 protein in a dose-dependent manner, with a near maximal response increase of 2.1-fold at 5 nmol/L. Leptin at 5 nmol/L significantly increased ACAT activity by 1.8-fold.

Leptin (5 nmol/L) increased the expression of the 2.8-kb and 3.6-kb ACAT-1 mRNA species by 1.6-fold and 1.7-fold, respectively, without significant changes in the expression of the 4.3-kb and 7.0-kb species.

3.2 Up-regulation of leptin receptor during differentiation of human monocytes into macrophages

Leptin receptor was revealed as a 150-kDa protein in Western blotting of human monocyte-macrophages. The expression of 150-kDa Ob-R was significantly increased from day 3 to day 7 during monocytic differentiation into macrophages.

4 Discussion

The results of the present study indicated that leptin increases acetyl-LDL-dependent cellular accumulation of cholesteryl ester in human monocyte-macrophages. This was explained by up-regulation of ACAT-1 at the protein, mRNA, and enzyme activity levels during differentiation of primary human monocytes into macrophages. Our results indicated that leptin at concentrations of 2–5 nmol/L, which could be found in the plasma of obese subjects, can enhance the expression of ACAT-1 and cellular

158

accumulation of cholesteryl ester. Thus, elevated leptin concentration seems to be a factor that stimulates the formation of macrophage-derived foam cells *in vivo* in humans.

In the present study, we first demonstrated time-dependent changes in the expression of Ob-R protein detected as a 150-kDa band by Western blotting during differentiation of primary cultured human monocytes into macrophages. The expression of Ob-R was up-regulated as monocytes differentiated into macrophages. Up-regulation of ACAT-1 by leptin may possibly be mediated through Ob-R.

5 Conclusions

We showed that leptin directly affects human monocyte-macrophages, up-regulating the expression of ACAT-1 *via* leptin receptors, and thereby promoting cholesteryl ester accumulation in these cells. Up-regulation of leptin receptor expression during differentiation of human monocytes into macrophages indicates stage-specific leptin signaling in human monocyte-macrophages.

6 References

Zhang Y, Proenca R, Maffei M, Barone M, Leopold L, Friedman JM (1994) Positional cloning of the mouse *obese* gene and its human homologue. Nature 372:425-432

Beltowski J (2006)
Leptin and atherosclerosis. Atherosclerosis 189:47-60

Glass CK, Witztum JL (2001)
Atherosclerosis. the road ahead. Cell 104:503-516

Miyazaki A, Sakashita N, Lee O, Takahashi K, Horiuchi S, Hakamata H, Morganelli PM, Chang CC, Chang TY (1998) Expression of ACAT-1 protein in human atherosclerotic lesions and cultured human monocytes-macrophages. Arterioscler Thromb Vasc Biol 18:1568-1574

Key word index